[illegible]in

'Somewhere [illegible] [illegible]au, Philip Judge pitches his radian[illegible] [illegible] a [illegible] about [illegible] and inscape, and is a luminous, funny and profound reading experien[illegible]

SEBASTIAN BARRY

'This is *The Good Life* meets *A Year in Provence* ... a lovely summer read. I thoroughly enjoyed *In Sight of Yellow Mountain*, a witty, warm-hearted and often delicious tale of transitioning from city life to low-fi economy living in the Irish countryside. It might be particularly interesting to the British post-Brexit brigade seeking a solution in the pastoral dream of the Irish countryside!'

SUE COLLINS

'*In Sight of Yellow Mountain* is part almanac and part memoir; a meditation on the meaning of life, the beauty of nature and the passing of the seasons, with added chutney recipes and tips on the best method for castrating rams. Acutely observed and beautifully written it is also erudite, entertaining and laugh-out-loud funny. As a portrait of family life and all its past trials and present joys, it is as honest as it is moving and utterly compelling. Never mind the bucolics, here's one man and his smallholding. A must-read.'

MARK O'HALLORAN

'*In Sight of Yellow Mountain* brought me back to my childhood love of James Herriot's books, *All Creatures Great and Small*. Philip Judge strips bare all dignity and personal embarrassment to immerse himself in country life and we are the winners. His anecdotes of encounters with local farmers and unsuccessful chutney fermenting will make you laugh out loud. But it also makes you want to put on your wellies and get out and stroll the Irish countryside to soak up the beautiful sights he so eloquently describes ... and if you're lucky enough, you might encounter some of the humorous characters and situations that he finds himself in! It is one you will not want to put down.'

PAT SHORTT

'An absolute joy to read. This book charms and enchants in a delightful way. A real treasure.'

DOMINI KEMP

In Sight of Yellow Mountain

A Year in the Irish Countryside

PHILIP JUDGE

Gill Books

Gill Books

Hume Avenue

Park West

Dublin 12

www.gillbooks.ie

Gill Books is an imprint of M.H. Gill and Co.

978 07171 7878 0

Designed and print origination by O'K Graphic Design, Dublin

Edited by Emma Dunne

Printed by TJ International, Cornwall

This book is typeset in Berkley Oldstyle with headings in ITC Benguiat

The paper used in this book comes from the wood pulp of managed forests. For every tree felled, at least one tree is planted, thereby renewing natural resources.

A CIP catalogue record for this book is available from the British Library.

5 4 3 2 1

Acknowledgements

Firstly I would like to thank Nicki Howard for that moment of elation over coffee in the Westbury Hotel when she told me she would like to publish this book 'just as it is'. I would also like to thank the rest of her team at Gill Books for making the journey to publication so smooth, particularly Sheila Armstrong for an astonishingly agreeable editorial process.

As to the writing, I would like to express my initial gratitude to Michelle Read, Karen Egan and especially Dillie Keane for their very heartening response to my earliest scribblings some years ago. In gathering my material I found Kevin Danaher's *The Year in Ireland* very helpful for its explanation of the origins of obscure country customs, some of which still residually survive. For the more practical agricultural knowledge and for more than a decade of hospitable friendship I thank Colin and Janet Tyner. As the manuscript took shape, Mark O'Halloran gave me invaluable reassurance that I was writing something more than self-indulgent whimsy. Audrey Brennan became an assiduous early reader and a sharp editorial ear as I completed the first draft. Her enthusiasm and benignly critical eye was extremely encouraging. Jane Karen read my first edit in two days and her response delighted me. Sebastian Barry and Alison Deegan gave me a reality check but flattered me by taking me seriously.

Completing the book was only the beginning of its journey to a publisher and many people helped along the way, including Eileen Fitzpatrick, Alison McGuire, Alison and Donnell Deeny, Karen Brade, Roberta Dunn and, vitally, Domini Kemp. Susan

Towers has provided much-needed social media advice as has David Quinn, along with some very useful scrutiny of my drawings and their position in the text.

For convivial companionship during many incidents recounted in the book I would like to mention someone who wishes to be acknowledged only as Roger Braintree.

And finally, I would like to thank someone without whom the book really would and could not have been written. She inspired it; she listened to every word; she suggested most of the changes and improvements; she steered the completed manuscript along its winding course to that meeting in the Westbury Hotel and, as in the book itself, she wishes to remain nameless.

Contents

PROLOGUE

The family computer is temporarily free of my small sons looking at cartoons of annoying oranges. I am sitting at it looking across the valley to Sliabh Buí – the yellow mountain. At this time of the rolling year, August, the name is apt. Golden cornfields predominate and, studded across them in pleasingly uniform ranks, the harvested bales are drying in the warm air. Interspersed amongst the yellow, there are meadows of light-green pasture.

Dotted with drowsy cattle, they make a geometric contrast, an oblique patchwork draped across the gently tilted hills – it is absurdly beautiful. It looks like an ad, but it is real. The harvest is being gathered. The sounds that reach my half-attentive ears are those of the full panoply of agricultural machinery. Combine harvesters lumber in the distance. From time to time tractors in low gear strain up the hill, towing heavy loads of grain or long trailers of hay and straw. Nearer by, the high note of a strimmer whines intermittently – a neighbour is trimming the verges of the fat hedgerows that line the usually quiet road on which we live. The country ants are at their harvest offices and I am happy but still slightly surprised to feel part of it all.

Twenty years ago in England, where I had been living since the age of nine, I imagined such a place at such a time but never foresaw actually inhabiting it. I was in a play set in County Donegal during the time of Lughnasa – the old Celtic festival celebrating the August harvesting. For a year, in London and on tour, I played a part I adored. My character spoke wry, haunting speeches of yearning, regret and misremembered boyhood and I spent those months lyrically evoking an Irish country childhood. In cities throughout the UK, in various theatres, through the year's cycle of seasons, the set remained the same: a small stone cottage amidst a swathe of ripe, golden corn. A decade later I found myself buying the reality.

This was not quite what I had envisaged when I was pacing the stage of a venerable theatre in the heart of the West End. Ambitious notions of theatrical glory hovered around the edges of my mind – London, New York, the World! Instead my geographical trajectory has been London, Dublin … a village in Wicklow. And the world, not surprisingly, is indifferent.

In retrospect, moving back to Ireland seems inevitable. Twenty-five years in England had not erased an essential sense of being from 'elsewhere'. Much of my work in London had been in Irish roles or Irish plays – an English education hadn't prevented my being seen as an Irish actor. I felt a little fraudulent, not having lived here since I was a boy, and I became adroit at changing the subject when asked in auditions what work I had done on the Irish stage. I suspected that singing 'Michelle, Ma Belle' for my granny in Glasnevin didn't count. Irish literary, musical and theatrical culture seemed to be dominating the world when I was in my early thirties and I was curious to get back to the source. To my mind, England had been getting greyer and grimmer, whilst, across the water, an inspirational Irish president, Mary Robinson, lit a candle in the window of her official residence to summon home the Irish diaspora. I accepted the invitation and then deferred it. I dithered and prevaricated for a year or two, slowly loosening my ties to London.

I allowed my career to languish (actor-speak for 'didn't work much'); although I did form a soul and blues band with a friend in order for him to play tasty guitar licks and for me to pretend I was Otis Redding. I had been meaning to do this since failing to be cast in an Irish movie about a Dublin soul band. I wanted to demonstrate that the casting directors had been fools to overlook me. We played a couple of pubs in London and did a triumphant tour of three clubs in south Wales. Having made my point, I was finally ready to move, and, apart from anything else, Ireland just felt like ... well, home.

I had been an urban dweller all my life. I was used to the lurid light of shopfronts reflected in wet pavements and weaving briskly through busy streets, dodging charity muggers. Capital cities have much in common, so when I left London for Dublin I quickly adapted and slept well. I added peaceful snores to the soothing noises of the streets – the steady growl of traffic, the occasional siren and the singing of jovial drunks. It was all quite familiar apart from the subtle difference in the songs that the revellers sang to proclaim their nationality – 'The Fields of Athenry' for the Irish and 'Wonderwall' for the English.

I met a cosmopolitan country girl – oddly, not a contradiction in Ireland – and she has been my beloved now for seventeen years. I knew of 'the country', of course. It was a place where one went for short but dull breaks to have your inappropriate footwear sniggered at by locals, and she didn't tell me much that changed my view. She had been raised in rural Leinster and when growing up was often entertained on family jaunts with gripping games of 'Name That Grain!' as they drove past yet another cornfield. On our occasional trips out of town she hooted with derision when I mistook barley for wheat, as I often did – deliberately, of course, just to amuse her. But our relationship grew in the bustling atmosphere of bars, cafes and restaurants, and we were both happy living in the city – until we decided to buy, that is.

At the time we were deafened and deranged by the roar of the Celtic Tiger economy (which has since reverted to the much more familiar mewl of a tentative tomcat) and we quickly realised we could hardly afford the smallest of apartments in town. I say apartment: we looked at a few tea chests fashioned from plasterboard and plywood and grew discouraged. Then the Beloved suggested we take our limited borrowing potential elsewhere. I was slow to catch on. She didn't mean crossing the river or exploring the suburbs: she meant out of the city and into the country. I laughed and then, seeing that look in her eye, stopped laughing and humoured her. I thought we'd maybe glance at a place or two and think better of it. It would be a day out. I would choose my footwear carefully and maybe amuse her with some cereal-related faux pas. We casually looked at two pretty houses and I felt grown up but not seriously so. One was isolated and the other needed work and commitment. Both prospects scared me, but a seed was sown, because they were cheap. The property bubble had not quite spread everywhere and at a certain distance from the city houses were almost affordable. We did some sums. Our tiny flat in town had a low rent and we had saved a little. Taking out a small mortgage on a place we could possibly let and occasionally visit might actually be feasible. We would never live there, of course, but a weekend retreat and a toehold on the property ladder seemed attractive and adult things to have.

The third house we viewed, and the first I really looked at, was suggested by the Beloved's mother. Until recent times, one of her preoccupations had been to explore every quiet lane in southeast Ireland in search of the perfect house: often making offers and just as often reconsidering and moving on. This barrage of bids over the years was largely responsible for the property boom. She has since settled – hence the recession. She looked at a cottage, discounted it, then thought it might interest us. We came, we viewed and, in a rush of blood to the head, we offered. The next day, to my alarmed surprise, the offer was accepted. A panicked call to our accountant assured us of mortgage approval by the end of the week. We had a deposit and my recently consistent earnings had seemingly reassured potential lenders of my financial probity, despite my dodgy profession – acting has never been an altogether respectable trade. That night after the show I was in at the time, a friend asked if I had done anything that day. I replied, still slightly stunned, 'I think I bought a house.' I've spent longer choosing a tie.

Then the wait began. We offered in spring and eventually got the keys as the harvest was drawing to a close. In the mean time I bought some wellies. There were delays involving land registry, as the house came with an acre. I was a landowner! I bought a tweed jacket. There were delays involving planning retention for the septic tank. I didn't know what that was. So I did some research and bought the Beloved some wellies. The spring dawdled into summer which ambled into autumn. And I really noticed the difference. We never had any real doubt that the house would be ours, so we made regular trips down to observe the changing seasons and admire the view from the front door across the fields to Sliabh Buí. We explored the area and discovered the nearby river flowing languorously through a forest of ancient oaks. We picnicked beneath the same trees in May amidst the bluebells. In June we found a fantastic pub on one side of a steep valley. We sat outside sipping our pints amongst the roses and felt we could almost reach over and pluck a sheep from the facing hill.

As for the cottage itself, I love it still. Although now it barely contains us two and a pair of large-spirited small boys, back then, at nearly a century old, it was in pretty good shape and mostly habitable apart from a vivid colour scheme which we rectified later. It also had an old stone piggery with its original trough. I was delighted by this discovery. I had always wanted pigs – or at least I imagined I had when I realised I owned a pigsty. I bought a tweed cap. I was now ready for anything and we took possession.

That first night I didn't sleep well. I couldn't relax to the nocturnal sounds of the country, which were anything but peaceful. Trees creaked loudly in the gentlest breezes. There was a startling cacophony of bleating when a fox worried the flock of neighbouring sheep. My fitful dreams were punctuated by the piercing squeaks and grunts of voles copulating in hedgerows. Dawn came absurdly early into our un-shuttered room but the gleaming sun illumined the valley below as it emerged from the morning haze. It was the time of year that Keats called the season of mists and mellow fruitfulness, and through the actual mists I glimpsed a possible future. I imagined vegetables ripening in my own patch and an orchard maturing with apples, plums and pears.

LUGHNASA

The Harvest

Picking Fruit

Ten years on, the abundant vegetables and orchard fruits I'd once only imagined are now manifest and I have more produce than I know what to do with. The open fireplace in our kitchen is packed with tiered jars of piccalilli, chutney, pickled shallots, crab-apple jelly, pickled cucumbers, elderberry cordial and sloe gin. I can't give it away. More accurately, I won't give it away. At least not until Christmas, and then only begrudgingly to recipients who will truly appreciate the gift. We went to Sunday lunch with some friends recently and, along with the obligatory bottle of plonk and the prearranged vegetable dish, we brought a jar of apple and courgette chutney. The response was polite but lukewarm. I was close to being offended until I saw, stretched along the kitchen window sill, ranks of green tomato chutney jars. At least, I gloated, they don't have as many varieties of chutney as we do!

Still, jars of produce have currency hereabouts. For a few successive winters some years back, we cut our Christmas trees from another neighbours' small plantation of them – in exchange for a jar of homemade mincemeat. This was a standing arrangement for a while, until the man died. The grief-stricken widow felled the remaining trees as they were too painful a reminder of her late husband. Or perhaps she just hated my mincemeat. Another neighbour who knows about plumbing (and is therefore a god) helped fix our pump in exchange for a jar of chutney … and a bottle of whiskey … and a few quid.

We watched a family movie together recently. The story told of a dwarf king whose lust for gold made his kindred delve ever deeper to satisfy his thirst for glittering riches. There was an image of him amidst mountains of sparkling treasure with a look of fierce delight in his avaricious eye. I look at my preserved hoard and feel the same. The thought of parting with any of it is hard to countenance and the memory of a recent loss of a large batch of pickled cucumber is still a source of emotional pain.

The jars in question were looking a little cloudy and emitting dribbles of thick, clear gloop – which is not ideal. We opened them and they were fizzing – which was not reassuring. Fermentation of some description seemed to be taking place, which, for an instant, was a little exciting. Maybe we had accidentally invented a new liqueur? I am open-minded and adventurous in trying alcoholic drinks. As an inquisitive teenager I experimented with many exuberant cocktails when raiding my parents' drink cupboard. As an adult, I have tried dubious and unique concoctions. However, mulched cucumbers in what looked like slug juice and smelt like horse piss was a step too far, and into the compost it all went. Half a dozen large jars of it. This was a tragic waste and, in terms of our very small-scale husbandry, an ecological disaster.

Talking of horse piss, the Beloved and I recently ate our way through a horse fart, which is the local name for a giant puffball. This has been a great year for mushrooms of all types (though we have confined ourselves to non-hallucinogenic varieties) and one recent misty morning, whilst gathering ordinary field mushrooms, Older Boy spotted what looked like two fungal footballs. We brought them home and looked in our book 'Mushrooms for Morons', which told us they were edible – though, just to be sure, we offered one to our farmer friends up the hill – thinking that if they ate it we'd be okay. 'A horse fart!' said Male Farmer Friend. 'I haven't had one of them in years. Lovely!'

He explained that when these particular mushrooms 'go over' they become yellowish and leathery to the touch – they also become hollow and when tapped sharply they split open, gushing a cloud of spores reminiscent of, to agricultural eyes, a horse's fart. This one, however, was soft and smooth and when cut it revealed itself to be made up of firm, white flesh throughout. Delicious fried like a steak, or curried, or in sandwiches, or in a risotto, it was big and took a lot of eating. The field mushrooms kept coming too. I have nearly, but not quite, tired of picking some first thing in the morning and frying them with an egg for breakfast. They have bolstered many casseroles and bolognese sauces. The freezer contains tubs of mushroom soup and cartons of sautéed mushrooms ready to do service. Sadly I have not found a recipe for mushroom chutney but I have found one for mushroom ketchup! Surely we need some of that in our hoard? 'No!' the Beloved yells. 'Enough with the mushrooms – please.' I don't understand this attitude. They are there for the picking … literally.

When confronted with a free bar at a function, I feel in some obscure way that I 'must get my money's worth' and drink it while it's going. I usually end up noticing that everyone else is talking quicker than I can think, or else I drag someone up to dance even though the music may have stopped half an hour before. It's a bit like that with mushrooms. Eat as many as you can while they're growing. Unfortunately, the ketchup recipe requires kilos and kilos and the fungal season is nearly over. But there is always next year.

Abundance is everywhere. Up behind the house, the vegetable garden, though late starting this year after a delayed spring, is still producing profusely. As we walk the boys down the hill to school in the morning, we pass a few hazel trees which are currently dropping their nuts on the path. I am keen to gather the lot but the boys insist I leave most of them for the squirrels. In the field behind us there is a large heavy-cropping crab-apple tree. Its fruit are deeply unpleasant to eat – I've tried – but mixed with a little sage or elderberry they make a delicious jelly which goes perfectly with roast pork. The blackberries in our hedgerows also make a fine jelly and this is the Beloved's preserve. We have a few remaining muslin squares we bought for wiping the boys when they were babies. She has recycled these into a complex contraption involving string and bowls. The pulped fruit drips slowly through the muslin. Then she extracts the essence and 'jellies' it. Despite frequent washing, these muslin squares are now much more deeply stained than ever they were when first used. This surprises me – I thought there was little outside an industrial laboratory that could match the potency of infant emissions.

And the orchard is finally bearing fruit. We planted three apple trees in the first year. More recently we added a plum tree, two pear trees and two cherry trees, and this summer just passed, we have seen a little action. Hopes have been raised then dashed. It's been a rollercoaster. As I write there is one pear hanging on one tree. This is very exciting. I had been warned: 'Pears for your heirs' – meaning, I suppose, that it would be years before they produced. But there it is – slightly scabbed, granite hard and singular. I took a picture of it and I admire it daily. It represents … I don't know, something or other: promise of future plenty; produce of past efforts; or paucity of the present, perhaps.

I look for metaphors and portents everywhere, particularly in the orchard and the night sky. I am a living embodiment of the observation that the danger with not believing in something is that you'll believe in anything. I see signs in many things. Although officially an atheist (albeit a Catholic one) I do occasionally feel the 'twitch on the thread' – the church might try to reel me back in at any time. I understand what Sartre meant when he said that he denied God's existence with every ounce of his intellect whilst yearning for him with every fibre of his being. It is comforting to believe there is purpose to and meaning in an apparently arbitrary universe. Acting work this summer has been, unlike the cow and sheep shit hereabouts, thin on the ground. Generally buffeted and alternately excited or disappointed by the uncertainties and vicissitudes of this absurd profession, it is steadying to ground myself in more tangible things such as shooting stars or what the fruit trees foretell. These all blossomed beautifully this year in a rolling wave of pale pink and delicate white, seemingly promising bounty. It looked like the cherries would crop especially heavily, which gladdened my heart. This indicated life was to be more than merely a 'bowl' of cherries: there would be armfuls, basketfuls – a cascade of riches! Then they all drooped and died. This meant that my hopes and dreams were insubstantial and inevitably doomed. Despondently I looked in my book 'Fruit Trees for Fuckwits' (most handy) and discovered that this was normal. Cherry trees do this as they are gaining maturity, building up reserves in preparation for fruiting. Ah! Good, I thought, this signifies merely a postponement of success – or at least a deferral of failure – maybe next year?

I turned my attention to the plum trees and watched carefully. Plenty formed like little hard, green almonds. Mindful of the message of the cherries, I was wary of being prematurely excited and waited patiently to see if they would ripen. They did! What a terrific omen! Surely plum roles were coming my way in profusion! I picked one and ate it. It tasted vile – sour, inedible and bitter with false promise. This meant that any plum parts that came my way would be unplayable – superficially luscious but integrally rubbish, leading inexorably to a humiliating end to my career. I regrouped and read the signs aright. When they were ready the plums would fall into my lap. 'Be prepared,' the orchard was saying – the readiness is all. Then most of the plums suddenly disappeared – some other bastard must have got the part! By the time I finally got to eat one or two they were practically prunes and that metaphor is too depressing to extend. There is still our one pear – lonely in its tree and far too vulnerable to bear the weight of my massive expectations.

At least we will definitely have cartloads of apples. All three trees have borne a heavy crop this year and, frankly, I can't be arsed to interpret this in any particular way – perhaps because the novelty value of picking and eating my own apples has now worn off. Besides, no apples taste as good as those stolen in childhood. I remember one particular tree in a garden near my granny's. Up a lane on the way to the shop presided over by a silver-haired man in a brown shop coat, there was a high wooden gate, and if you shimmied up it (barking your short-trousered knees on the flaking paint) and heaved yourself high enough to peer over the top, you saw it – a solitary, spreading Granny Smith tree casting a cool shadow in the warm sun. Festooned with ripe, inviting apples winking in the golden light, it was utterly irresistible. In later years, when trudging through Milton's megalithic *Paradise Lost* and reading of Eve's yielding to Satan's honeyed words, I came across the line 'Such delight till then, as seemed in fruit she never tasted', and I remembered the first apple I stole from that tree. It was luminous green, crunchy, crisp and utterly delicious. The slight sharp tanginess concentrated the intense sweetness. And it was huge. It took two hands to hold it steady as I gorged myself. As a young adult, I often regretted that you couldn't get apples as big as they had been in Ireland when I was a child.

Eventually I realised that it was I who had grown and not the fruit that had shrunk, but nothing subsequently has lived up to 'the scent of that alluring fruit'.

Our earliest maturing tree – Beauty of Bath is the variety – produces small, sweet apples that are crisp for a day or two before turning to cotton wool. They have to be dealt with swiftly. This year I borrowed a juicer and extracted a couple of litre bottles of pale pink nectar. Quite delicious once you had skimmed the acrid, frothy scum from the top – maybe I should have peeled off the brown scabby bits first. The Beloved also made two tarte tatins from the crop, but most became food for wasps. They burrowed into the apples whilst they were still on the tree, eating them from the inside out and carving a cavern within. If you picked one, as many as half a dozen of the thieving yellow-and-black bastards could emerge from one small hole in the skin. Once I got over my initial rage, I found it gratifying to pluck an apple and fire it as far as I could into the neighbouring child-free field. Watching the dizzy wasps hurriedly escaping from the flying fruit in drunken pirouettes reminiscent of a comet's trail is most amusing. That tree is now done for the year, and as for the other two, which are also threatening abundance ... well, we might get rid of some at Halloween.

The harvest is nearly over. Our swallows have gathered, twittered and gone. And the noise of combines and tractors is dwindling. Before the advent of modern agricultural machinery, when crops were reaped manually, all hands would gather in the field to watch the last stand of corn being cut. Enough for a sheaf would have been set aside for this closing ceremony and, because the retreating animals would have stayed under cover for as long as possible, this last stand could shelter frogs, corncrakes, partridges and sometimes even a hare, who would hide till the final flash of the sickle. Finishing the harvest gave rise to the phrase 'putting the hare out of the corn', and if you wanted to mildly gibe someone who was a little late with their harvest, you would say, 'We sent you the hare!' When I discovered this, I wished I grew wheat or barley. 'Putting the slug out of the lettuce' really doesn't sound as good. Still, it's nice to walk in the cornfield opposite us at the end of a warm day and stroll among the last rows of huge rolled, golden bales as they cast their lengthening shadows across the yellow stubble – 'a dusk El Dorado', Heaney calls it. The boys love clambering up on top of them and lying flat on their backs to survey the still sky. Seeing them lying there looking up chimes with a vague memory I have of doing the same – though I can't remember where or when. A race memory, perhaps: children have been lying on mounds of straw pondering the wheeling heavens for centuries.

Here in Ireland, within living memory, at this time of the year young people gathered on higher ground and hilltops for games and sport and bonfires to mark the end of Lughnasa. The old legends tell how Tailtu, foster mother to the sun god Lugh, died of exhaustion after clearing the central plains of Ireland in readiness for cultivation, and that Lugh thereafter held funeral games in her honour at harvest's end. These seasonal customs survived even into the middle of the last century. (This is not so very long ago. I was, after all, born round about then – much to my sons' astonishment: the 'olden days' they call it.) The highpoint to these games was when the young men leapt over the bonfires to prove their prowess – an urge I dimly recall myself and still see exemplified in my boys. The other evening I lit a bonfire of hedge clippings. I did it the traditional way, using a complete box of matches and a pint of petrol. The two of them leapt like young bucks through the smoke as it rolled in dark billows up the hill. Then the wind changed and it rolled in darker billows down the hill and across the field, fumigating our neighbours. I owe them a jar of chutney.

Another harvest tradition was throwing whole ears of corn into the fire to cook them quickly. The straw and chaff would burn off whilst cooking the grains within. The grain was then dried and the remaining ashy chaff was winnowed by the wind outdoors. This was done on the morning of Lá Lughnasa (the first day of the harvest) to make bread to eat that evening. The custom became so common and general that in 1634 the Dublin Parliament passed 'An Act to Prevent the Unprofitable Custom of Burning the Corn in the Straw'. The practice seems far too much like hard work to me so I just toast some marshmallows on a stick and the boys are happy. This is a ritual they now expect after any bonfire or barbecue and it is a custom I am happy to observe. Their expectations now dictate more than one of our seasonal family customs. For example, we are now regular attendees at and contributors to the Protestant church's Harvest Festival. This is a notion my younger self would have shuddered, or at least sniggered, at.

Neither I nor the Beloved are believers. We were both baptised Catholic but it is years, decades, since either of us engaged with what Larkin called 'that vast, moth-eaten, musical brocade'. I would best describe myself as a lapsed dialectical materialist and the Beloved believes in brisk, no-nonsense 'living in the moment', with a little low-level witchcraft thrown in. There isn't an educational establishment in the village that actively promulgates these principles – just the two usual Christian dispensations are catered for. We chose the Church of Ireland school because it was nearer – only a short walk down the hill in the morning; it was also smaller – two teachers for about thirty kids; and the children seemed happy. Besides, I thought the religious instruction would not be too dogmatic. Having grown up in England, I had a notion that not believing in the divinity of Christ was no bar to advancement in the Anglican Communion. As it happens, the school is a touch more devout than I thought, but so what? If the Older Boy talks of God, I chat about Allah, Yahweh and Zeus. As for Younger Boy, well, he recently described his concept of the divinity to his mother: 'God is a giant adult who is bigger than the Hulk and he looks like a Viking and is probably a bald angel.' He added that he believed in him and prayed to him – but that it didn't work.

So we go to the Harvest Festival and I bring a basket of vegetables to adorn the altar and we both smile tearfully at our boys singing their hearts out. We stay quiet, though. No joining in with the responses – just to demonstrate what radical free-thinkers we are. I doubt I could join in even if I wanted to out of politeness. It would remind me of the dreary monotony of the Catholic liturgy blandly mumbled in a cold, depressing concrete church in the industrial town of my grey teenage years … or so I thought.

At the most recent Harvest Festival the church found itself slightly short-handed, especially among the ranks of the choir. We are still blow-ins, of course, and always will be. But our boys are locals, which makes us semi-established. It is also not unknown that we have both trodden the boards whenever allowed to. So we were dragged up to sit in the top pews with the top Prods of the village on condition that we sang out good and *loud*. Tunefulness didn't seem to matter. I tried to wriggle out by protesting ignorance, but my excuse of not knowing the hymns was swatted aside and I surrendered. Volume not melody was the key, they said. I know how to take direction so I bellowed and thoroughly enjoyed myself. I haven't worked much recently, and besides, youthful radicalism is eroded with the passage of time – and I am now middle-aged after all.

SWEET GERANIUM APPLE JELLY

Ingredients

2.5kg crab apples or windfall cooking apples
2.5 litres of water
2 unwaxed lemons
Sugar
10 large sweet geranium leaves

Method

Wash the apples and cut them into quarters. Be sure to cut out any bruised bits from the windfalls. Put the apples in a large saucepan with the water, the thinly pared zest of the lemons and the sweet geranium leaves. If you don't have a sweet geranium, this recipe is also delicious using sage or rosemary.

Cook until reduced to a pulp – about half an hour. Spoon the pulp into a jelly bag or tie it up in a piece of muslin and allow to drip until all the juice has been extracted. I leave mine overnight. Don't be tempted to squeeze the jelly bag to get out every delicious drop – it will make your jelly cloudy. Measure the juice into a preserving pan or large saucepan and allow 450g sugar to every 600ml of juice. Warm the sugar in a low oven.

Squeeze the lemons, strain the juice and add it to the pan. Bring to the boil and add the warm sugar. Stir over a gentle heat until the sugar is dissolved. Increase the heat and boil rapidly without stirring for about 8–10 minutes. Skim, test and pot immediately. You can put a fresh sweet geranium leaf into each jar as you pot the jelly.

Herding Cattle

Many years ago in London, when I was younger and less yielding, I acquired a coat I had yearned for. It was one of those Australian stockman things – black, waxed, nearly floor-length and with a little half-shoulder cape: Trail Dusters, I think they were called. I loved it. I sashayed around Soho delighted with myself. I thought I looked darkly cool, with an air of edgy, understated flamboyance. I now realise that the effect was closer to a shabby, slightly inebriated Christian Brother. I kept the coat, though. It reminds me that, although I have never been a fashion victim, I have occasionally been a sartorial fool. It had hung for a year or two, half-forgotten, on the back of the door to what was then our spare room. I didn't quite have the courage to wear it outside for fear of looking like an over-dressed, over-keen eejit – which is what I often am. One day Male Farmer Friend asked if I'd help him herding cattle for the day and of course I said yes – I had the very garment!

The next morning I strode resolutely up to the 'cross' – or the crossroads at the top of our hill where the farm is. I was dressed to impress in wellies, the coat and a jaunty black waxed cap to match. I also carried a stick – for waving at cattle. As I walked, I rehearsed useful phrases such as: 'Hup!' 'Easy there!' and 'Go on, you good thing!' to throw out casually when necessary. I arrived at the barn and Male Farmer Friend tactfully suppressed a grin. He said he hoped I didn't mind getting my coat spattered with cow shit. My hesitation was only split second. 'Not at all,' I replied cheerfully. It hadn't really occurred to me that I would get crapped on. I thought I would be gently encouraging Daisy and Buttercup along a quiet lane with a calm but firm click of my tongue and the judicious use of my rehearsed useful phrases.

We headed down to the 'Long Field', Male Farmer Friend driving a quad bike and me riding precarious pillion – the long flapping tails of my (now evidently impractical) coat threatening to whip me under the wheels. Relieved to get off, I was positioned behind a large beech tree and told the plan. It was simple, really: the thirty-odd head of cattle would be rounded up and politely encouraged out of the field with kind words delivered from the quad bike; their inclination would lead them down the lane instead of up, which was actually where they were bound. When they were all out and the gate closed behind them, a whistle would be my signal to emerge from my station, stop the herd and send them back up the hill to the barn. I was pretty clear on most of this, though slightly vague on the me-stopping-the-herd bit, but before I could demur, away the quad roared and I was left alone and slightly anxious.

A morbid silence settled. The sun shone pitilessly. The shadow of the beech tree stretched before me like a gallows. Life seemed intensely sweet and far too short. I drifted into a mournful reverie … my mangled, cattle-stomped corpse … weeping friends, inconsolable family, torch-lit procession following my hearse through the darkened village … distraught Male Farmer Friend racked with guilt. I was startled by the raucous caw of a malevolent raven and came to. Behind the beech tree, I was suddenly aware of many confused sounds. The air was thick with roared obscenities and the revs of the quad bike. The ground rumbled with the urgent clatter of heavy hooves. There was a loud whistle, then a shrill one. Then a bellow of 'Stop the fucking cows!' My doom was upon me. I stepped out, looked up the lane and narrowly avoided fainting.

A single cow trotting is quite a pleasing sight. It is a gratifying combination of bulk and elegance. Imagine a hippo walking a tightrope or a large naked lady skipping to the bathroom in kitten-heeled slippers. However, a whole herd cantering straight at you is a different matter. I made a strange, involuntary whinnying sound. Then I took hold of myself and uttered a commanding clicking noise. Nothing. I cleared my throat aggressively and tried a rehearsed useful phrase or two. On they came. There was nothing for it but to deploy the coat. I opened it wide and pointed it straight at the lead cow, sneered, waved my arms frenetically and went '*Rooaarraarrggghawhooeerrfft!*' The herd stopped in its tracks and stared timidly at me. I snorted with relief and waved my stick in delight. Then, with a nonchalant air that belied my inner confusion, I herded them up the hill.

The coat had not let me down. Possibly what the cows had seen was not the actual tremulous pillock in the inappropriate outfit, but an apparition of an outraged, thundering priest ranting of damnation and hellfire (they were Irish cattle, after all) and it had stunned them into submission. Or else they thought they saw a mad, giant crow and it really messed with their heads. Either way it did the trick. Meanwhile, back in the barn, a white-coated vet was testing the herd for TB and brucellosis (a bacterial disease affecting cattle) and 'the ladies ' – as I was now calling the cows, with all the authority of an ingénue – had to be encouraged into a narrow bottle-necked passage called a 'crush'. Here they were held still long enough for the vet to do his thing and for them to defecate copiously, mainly over me. I didn't mind that much – the coat took most of it. Besides, they may soil me but I eat them, which seems fair. Mysteriously, the vet's white coat remained unbesmirched throughout the day. I cannot say the same for mine.

I was stockman again for a day or two more recently and, being now an old hand at driving cattle, I didn't feel the need to dress with attitude so I left the coat at home. Amongst many other agricultural outfits, I have a brown leather jerkin I mostly use for gardening. It is supple, faded and judiciously stained from long use and it has many pockets for penknives, bits of twine and small, useful things. It declares its owner to be relaxed, practical, ready for anything and a little rugged. I put it on over a worn pullover and old jeans and ambled easily up to the cross.

We gathered the cows from various fields without any noteworthy events and put them in the barn. They then had to be brought, a dozen or so at a time, into a holding pen prior to being driven in single file through the crush to allow the vet to administer the jabs for the TB test. I hopped into the holding pen with them, the easier to encourage them into the crush, and, feeling quite relaxed and confident in my herding skills, I began to pay attention to what the vet was actually doing. I was leaning on the gate, watching intently, when I found myself suddenly much closer to him – about four feet closer, and indeed three feet higher. It seemed a playful cow had got her head squarely under my buttocks and hoicked me into the air just for the craic. I was a little shaken, but essentially undamaged. I also figured I had 'made my bones' and was now a fully initiated stockman, but the vet was scornfully amused when I asked if I had just been 'blooded'. He replied with a curt shake of the head and a slightly derisive grin. Only a mite disappointed, I turned my attention back to where it should have stayed – with the cattle in the pen. They were doing their thing: bellowing, urinating and defecating. And I was in there hugger mugger with them – their orifices around chest height and mostly aimed at me. It was a cloacal Jacuzzi with jets of piss and shit squirting from random directions. At one point I was a little panicked to find myself face to face with the bull – he had a ring in his nose and everything. For an instant I froze, but he just wandered placidly past me into the crush like a good boy. He had, after all, recently impregnated an awful lot of cattle so perhaps he was feeling relaxed. The concrete floor of the holding pen was now growing decidedly slippery, and we were all slaloming around in the slather and steam when a cow 'accidentally' trod on my foot. Luckily she only nipped the edge of a toe but I was wearing wellies without steel toecaps and, oy vey, it hurt. I looked at the vet for affirmation: 'Now am I blooded?' All I heard was 'Hah!'

I limped home disconsolately: shite-spattered and exuding a sweet, sharp stink – a quintessence of grass. My leather jerkin was now much more comprehensively stained. When my sons saw me they laughed with savage glee. The discovery that I had cow poo in my pockets was particularly hilarious. That night they were still sniggering in their sleep.

Two days later I was called upon again. The TB test consists of a pair of jabs in each animal which are later examined for comparative inflammation. It is a way of tracking and containing the disease if it occurs rather than a means of curing it. So we had to go through the whole procedure again. This time I thought it wise to redeploy the coat. I got it out of the garage and examined it. It had long since lost its waxiness but it now had a much more effective barrier – a complete carapace of dried-in cow manure. It was an all-in-one suit of armour. I strapped it on and walked stiffly but confidently up to the cross. This time the cows were much more reluctant to move, it having only been a couple of days since they had been 'persuaded' into the crush and jabbed twice. They were consequently in an obstreperous mood. One in particular, with an evil leer and an ugly disposition, stopped in the gateway coming out of the field, stared at me gimlet-eyed, lowered her head, snorted and pawed the ground in the threatening manner of an enraged cartoon bull. I admit that I quailed, but I had to tough it out, so I waved the coat and the cow backed down.

Relieved but a little disconcerted, I followed the herd up the hill. Getting them into the barn took a lot of shouting and roaring and a fair amount of whacking bovine ass. Decanting a dozen into the holding pen was proving impossible – they kept moving down into it and then wheeling back up and out before we could close the gate. You really don't want to close a gate in the face of advancing cattle – you will be pasted. You let them out and try again. I was just stepping back to allow them past me after the third or fourth attempt to shoot the bolt behind them, and I was on the point of springing forward to secure a few who had loitered in the pen, when one of them made a leap for freedom. She practically vaulted – all sixteen hundred pounds of her. It was awkward, ungainly and strangely beautiful – almost in slow motion, it seemed. And as she passed me in mid-flight, she sort of swerved her body, twisting her spine and rotating her pelvis, and kicked me hard on the hip.

I was batted back against the wall, shocked and winded, but I managed to pull myself over and out of harm's way. I gathered my wits and felt for damage. The coat was a protective shell but even it had its limits. There appeared to be nothing broken but I felt a general sense of numb elasticity in my hip and a dull pain spreading around it. It was time for mild heroism. I girded up my coat again and returned to the fray. Of course, in the general melee, Male Farmer Friend and the vet hadn't noticed this rather momentous event. And when the cattle were in due course secured and things had calmed down again, I mentioned with, I thought, quite noble understatement that I had been kicked. Hard. But that I would be alright to continue in a minute or so. I thought I was being really very brave. I got a couple of sympathetic nods. And that was it. No concerned solicitousness, no medal, no nothing. I asked the vet again did this count? Was I finally initiated? And again he shrugged and shook his head. I don't know what he wants. Do I have to be anally gored before my hazing is complete?

After the vet had left, having established there was no TB in this herd, we were sorting the cattle in readiness for bringing them back to their respective fields when Female Farmer Friend got a kick from the same cow. I now recognised her as the one who had snorted at me at the beginning of the day. Possibly she was also the one who had butted me in the arse in the holding pen. She must have taken some evolutionary step and was having great sport twatting bipeds about the place. We must eat her soon, I thought, before she spreads her poisonous, seditious ways.

Female Farmer Friend, meanwhile, having taken the blow, had been flung to the floor and was sliding ten feet through slurry till the barn door stopped her. The cow had really connected – she must have got her eye in. Anyway, up she got (Female Farmer Friend that is, not Kicking Cow – that beast was now skulking in the herd). She shook herself, surveyed her totally-soiled-from-head-to-foot overalls and hobbled towards the farmhouse. Her husband nodded and said nothing but watched her walk away with a proud glint in his eye. His steady gaze followed her through the door and into the house, then he turned his stern brow to the herd and fixed the culprit with a grave look. I have known these people for ten years now and I can testify to a happy, loving marriage with a deep and easy affection. But there is no place for sentimentality on the cattle field. For uncomplaining fortitude, the oft-caricatured English stiff upper lip is merely a pale shadow of the grim forbearance of Irish cow folk. After this the cattle were, well, cowed. They went submissively where they were bid and then I went home and spoke no more of my injury. What vengeance, if any, Male Farmer Friend wreaked on the miscreant heifer I do not know. Whereof we cannot speak, thereof we must pass over in silence, as Wittgenstein once said in similar circumstances.

Sawing Wood

The days are getting shorter and the leaves are falling. It will soon be time to light the stove again. I am proud and fond of our stove. It stands in a corner of the living room on a little plinth that I built and it has warmed us and belched occasional smoke at us for years now. When we first bought the house we found that we spent very little time in the sitting room. It was a still and quiet space, despite the loud pink on the walls – which we have since painted over. It had a nice view down our acre to the old oak forest across the valley. But I found if I sat there reading the paper, the pages started wilting after a few minutes.

The cottage is a century old, but it has central heating and double glazing – the problem was in the much more modern extension out the back: a single storey tack-on hastily knocked up out of breeze blocks and damp as bejaysus. Half of it consists of the bathroom, which is tiled fore and aft and still seems watertight. The other half is the aforementioned sitting room, where we didn't really feel like sitting very much. Something had to be done. The Beloved looked in my direction … Ah! Something had to be done by me.

I had recently begun to get acquainted with my inner handyman. I was rather surprised to find that he existed at all, never having been particularly practical. I quite liked Lego as a child but Meccano scared me. I suppose I must have changed a plug once or twice without electrocuting anyone. Oh wait, there was that time when … but drink had been taken and the relationship was doomed anyway. However, once I actually owned somewhere, I found I was keen to 'fix' things. I was still feeling quite smug about having replaced a rotten window sill in our bedroom, even if the replastering was more jazz than classical. And it was still a pleasant novelty to me to be hanging out in hardware stores talking of tools. So I cheerfully set about fixing the sitting room. Off I went to have deep discussions, take advice and buy stuff. Within not too long a time, and to my smug satisfaction and the Beloved's surprise, I had dry-lined the external walls, put in skirting boards and built a raised concrete platform to receive the stove. I was thrilled with myself and I still occasionally pat the plinth to reassure myself of its enduring solidity.

The stove has done great work over the years but it does smoke us out from time to time. Sometimes because of a sudden veering of the wind which sends the smoke back down the flue to mushroom out into the room – eliciting excited yells from the boys; and once or twice because we have forgotten to clean the flue and it has become completely blocked – which is pretty bloody stupid of us. The result was a steady greying of the nice off-white colour scheme we chose. The recent drawing-in of the evenings emphasised the gradually accumulating gloom of the room so we repainted it. The Beloved did the ceiling and I did the walls in stages, moving furniture, dinosaurs and the boys' arsenal as I went. When finished, we stood back for an admiring appraisal and the customary exchange of congratulations and compliments on another job completed in readiness for winter.

The stove is now lit and, as I write, the flickering orange glow keeps drawing my eye from the laptop. There was a light frost during the night, and when I opened the shutters this morning the grass glinted slightly in the field. I went out to the woodpile in pyjamas and Crocs. The air was clear, crisp and scrotum-tightening. I filled the baskets with logs, sticks and kindling. I have it all arranged in the piggery, which is currently full. Like the store of chutneys and pickles, a large stockpile of fuel feels like a bulwark against the cold months to come. We burn a lot of wood and I am often busy gathering it. We try to limit our use of fossil fuels. Like most people we are aware of the ecological cost of peat and coal and they are also much messier. Coal burns bright and hot, though. In a particularly cold winter it can really bang out the heat and we do use it from time to time – we are not fundamentalists. However, it leaves dirty black ash and a stony residue – clinker, as I have learnt to call it – which needs to be shovelled out each morning. This feels industrial and urban rather than agricultural and rural and therefore a little grubby.

I know I am conveniently ignoring the manner in which most of the electricity we consume is produced, but 'hypocrisy is the homage vice pays to virtue' and true ecological virtue would require us to heat the house either by composting our hair clippings or with vigorous exercise on a treadmill linked to the mains. Male pattern baldness would render the one, and middle-aged indolence the other, ineffective. We sometimes burn peat or turf, and I do like the smell. It has a Proustian effect. It summons memories of childhood comfort and warmth or pubs in the west of Ireland – adult comfort and warmth, I suppose. But the peat bogs are being steadily eroded, which is something we should try to halt, and besides, a turf fire produces a very fine, all-pervading beige dust which settles everywhere. The old Irish peasantry were remarkably resistant to the colds and flus usually associated with a wet, cold climate such as ours. I suspect this had less to do with innate hardiness and more to do with the fact that the average Irish lung was coated with a thin protective barrier of turf ash.

So we burn a lot of wood. And I have to gather it. Fortunately we are surrounded by tree-lined fields which shed branches, boughs and even the occasional whole tree when the wind is high. Male Farmer Friend often mentions a bough of an oak or a small ash – and once an old beech – which may have come down. Fallen timber can obscure the cattle's path to water and prevent the tractor from ploughing or sowing. This mutually beneficial arrangement means he gets his field cleared and I get the firewood.

This only became possible when I got a good chainsaw. I was equally excited and timorous when the Beloved got me my first one for Christmas a few years ago. The history of our reciprocal birthday and Christmas gifts records how our lives have changed over the last decade. I used to buy her lingerie, dresses, jewellery, books, perfume and (affordable) objets d'art – things calculated to delight her eye, amuse her mind or adorn her body. She would buy me similarly lovely stuff – fine wines and good shirts, interesting music and, on one occasion, a nude portrait of herself, for which she had been secretly posing for months. When I hung this last gift in the house I rented at the time, some friends found it slightly discomfiting. Perhaps they thought it was a kind of hunter's trophy – I have always thought it a really cool present. It now hangs partially obscured by accumulated junk in the 'spare' room, biding its time until it can embarrass our sons when they reach adolescence.

I realised the nature of our gift giving had fundamentally changed when one subsequent festive morning I was thrilled to get an electric sander. In that same exchange I melted her heart with a large stockpot. I admit that a lot of the 'really useful stuff' we have recently bought each other regrettably conforms to traditional sexual stereotypes, but these are hard to avoid. We inhabit each other's space a lot, given the small size of our cottage and the ever-growing size of the boys. Compromise and give and take are not only desirable: they are necessary. Consequently most decisions are jointly made and all *Lebensraum* is shared … except that somehow, by some unspoken moral force, the kitchen is slightly more her domain than mine and I am largely lord of the garage.

But we also play less traditional roles. I am the undisputed master of laundry and she is the go-to parent when the boys have any toy-construction problems. To see her recently build some tricky First World War model biplanes I had bought them was to observe a masterclass. I remember a time in our first year here whilst I was at the Lyric Theatre in Belfast, playing an LA film producer in a Sam Shepard play. It was the cold, blustery month of March and my pastry-pale northern European skin needed assistance to appear convincingly Californian. Meanwhile, the Beloved had been hacking, chopping and clearing our field. I rang her to ask how her day had been: 'Great,' she responded, 'I've been felling trees on our acre! What have you been up to?' I was pleased to reply, 'I've been having a spray-on tan in a beauty parlour.'

That first chainsaw was by no means top of the range, coming as it did from one of those low-price German supermarkets – Adl, was it? This meant that it didn't last very long. I got about a year's sporadic use out of it before it seized up and died, never to be revived. Its repair was beyond even a mechanically gifted neighbour who can strip a lawnmower and use the parts to fix a tumble-drier. But during its short life it taught me the basics of chainsaw etiquette, which can be simplified as BVFC (Be Very Fucking Careful). I had a really healthy respect for the machine, by which I mean it scared me witless. But I had casually mentioned in the Beloved's hearing how useful a chainsaw might prove, and now that I had one, I was obliged to use it. I had to psych myself up in order to start it. I would spend a day or two mentally preparing by frequently eyeing the doomed log and thinking, 'You're for it!' I would also throw out random remarks like, 'I must saw that wood.' This was done repeatedly until eventually the accumulation of empty words had to be substantiated with action and I would go and get the chainsaw.

It always took quite a few hard yanks before it roared into loud, scary life. Then I would grip it determinedly and set to, whilst inwardly whimpering, 'Yikes!' When it worked, it really worked, ripping through wood as quickly as it could doubtlessly rip through your thigh, and I was always relieved to put it away. When it finally stalled forever I was secretly happy to leave it behind me. However, I had awoken my inner lumberjack and, whilst he didn't want to work all day, he wouldn't go back to sleep.

I have now conjured up a handful of inner personae, which is partial compensation for a lack of good acting roles. It amuses me and the Beloved indulges the whim to the extent that she tends to give these characters names, thus calling them into more tangible existence. The first one she recognised, christened and confirmed was my inner gardener – this was after we planted a few vegetable seeds in our first year and I took the responsibility for tending them. She calls him Mellors, after Lady Chatterley's lover, the eponymous lusty gamekeeper in the novel by Lawrence. This achieves two things to her purpose – it casts her as the elegant chatelaine of a substantial house and it indicates to me certain robust services she expects. My laundering skivvy she named Perkins and my inner handyman is called Quigley. Actually this last character was named by the *soigné* maître d' of a restaurant we used to frequent in Dublin and I don't really want to explain why.

Anyway, inner lumberjack was waiting in the wings, eager to get back on-stage. So just over a year ago, when the autumn was turning the trees red and gold and the woodpile was little more than twigs and sawdust, I bought a new chainsaw. More accurately, my credit card company did, and as Christmas was on the horizon and my mum was keen to take advantage of a whole new field of present-buying opportunities, she got me the protective gear. After all, every character needs a costume. Apart from poor old Perkins, that is, who, like any drudge, will wear what he's told.

The safety outfit consists of extremely tough trousers, saw-proof gloves and a helmet with attached visor and ear protectors. This ensemble makes me feel quite butch – in a Village People kind of way – and also very safe. I intend to always wear it and never need it. It is especially necessary when I am amidst the tangled branches of a fallen tree and sweatily trying to reduce large boughs to more manageable proportions. As they say around here, getting firewood 'warms you twice – once when you're cutting it and once when you're burning it'. Lumberjack has yet to be christened. Perhaps this is appropriate: he is probably self-contained, self-reliant, taciturn and mysteriously nameless. The Beloved suggests he might be called something embarrassing like Lesley and just doesn't want to talk about it. Anyway, whatever his name, he has been busy throughout the summer and the piggery is crammed with wood.

Raising Rams

I won't soon forget the morning I castrated Donald and Milo. Someone was going to have to do it and the Beloved felt that 'someone' ought to be me. I hadn't expected that I would have to tackle it with a hangover, though, and when Female Farmer Friend rang to summon me to 'do the job', my stomach lurched queasily. Donald and Milo are, of course, two young rams. We had hand-reared them since they were a few weeks old and we knew this day was going to come at some point.

Up at the farm, there are a few orphaned lambs to be dealt with every year. If a ewe (or 'yo' as they are called locally) has three or four lambs in a litter, which is not uncommon, she may reject one for no obvious reason. If a ewe dies in labour, which is rare but not unknown, then her lambs will need tending. Sometimes you can pass a lamb off on another mother if she has recently lost one of her own. In a large litter there is often a runt which doesn't survive. When this happens, you skin the dead lamb quickly, ideally while it's still warm, and tie the pelt to an orphan or rejected lamb that needs to be suckled. I've seen it done and it looks quite bizarre: like a baby lamb going to a fancy dress party … as a dead baby lamb. The smell of the tied-on skin persuades the ewe that the lamb is actually hers. This usually works – sheep are really not the most perceptive of animals – so with a bit of juggling and ovine social engineering, most lambs can be placed with new families. I am all for adoption really.

However, there will always be an unplaced few that need bottle feeding, and this year the Farmer Friends suggested we take a couple off their hands to rear and, ultimately, eat ourselves. This was an attractive proposition. We imagined future delicious dinners during which the question 'Where did this lovely lamb come from?' would be answered by us casually glancing out the window and saying, 'Just there, actually.' In this scenario we would quickly move the conversation on to the horrors of our overflowing septic tank lest we appeared sickeningly smug. It also made sound financial sense to rear a pair of lambs ourselves if you didn't factor in our labour. But we're actors – we're cheap. There would be a certain cost in grain feed and lamb formula milk, but sheep also eat grass – we had plenty of that, and I could knock up a pen for them and move them around the acre. There would be an added expense because every animal in the country is numbered due to food traceability legislation, and when the time came to be slaughtered they would need to rejoin the flock in order to be 'processed', which would incur a small fee per lamb. We would also have to take our ones separately to a butcher's to be jointed for the freezer – another few quid. But the total cost was still going to be far, far less than we would spend on the equivalent amount of cutlets, legs and shoulders. So we told the boys that some lambs were coming to stay and we deliberately avoided using the word 'pet'.

They arrived with names, which I am told is not a good idea, although, personally, I don't see why you shouldn't name your food. Dinner or Derek – I don't see any difference, really. You are either going to get attached to an animal or you are not, whether you refer to it as 'Lucky' or 'Number 42'.Farmer Friends' youngest daughter still weeps at the words 'Big sixteen.' As a child, she became very affectionate towards a particular sheep. It had the number sixteen sprayed on its fleece to indicate which ewe was its mother. It was also slightly larger than its siblings – hence Big Sixteen. In due course the particular sheep went off to the 'factory,' as the girl had seen countless others do before. She is now a capable, independent, city-dwelling young woman but she still wells up at those words.

'A rose by any other name would smell as sweet' as Shakespeare has it. But then 'A rose is a rose is a rose!' as Gertrude Stein had it, and she has a point – a lamb is a lamb is a lamb. And we now had two called Milo and Donald. They were named by Belgian acquaintances of the Farmer Friends who are regular holiday visitors. They sprayed the letters LO after the M1 which already been sprayed on the fleece of one of our lambs as an identification mark. This spelt the name 'Milo', which they thought a hoot – Belgian humour, I suppose. The other lamb had a slight underbite which they thought gave its mouth the appearance of a duck's bill, thus 'Donald'.

Our sons were ecstatic to have them and feeding them was fun. They almost knocked the bottle from your hand as they head-butted towards you to feed. And they suckled like industrial pumps, with much excited slavering and mess, which caused the boys huge hilarity. As they got older and bigger they became able to leap over their pen, but they only really did this at the prospect of being fed. The sight of them springing over the top and running eagerly towards you every morning as you brought them their food was very endearing. Occasionally over the summer we let them out to roam the field for the day and our sons played with them. To see the boys running around and the lambs kicking up their back legs and actually gambolling across the grass after them was lovely. But I never became sentimentally attached to them. I never forgot that they were destined to be dinner. If I had named them I would have chosen Korma and Kofte. However, we were unsure of what exactly to tell the boys.

As the lambs became more confident of their ability to escape from their pen and wander the field, they started to range further, and one day Donald nipped out the open upper gate and off down the road. I was alerted by a child's howl of abandonment and sprinted after him. After fifteen yards I slowed to a trot, partly to avoid scaring the lamb further down the road and partly to avoid collapse, and I heard Younger Boy sob despairingly, 'Don't run away, Donald!' It was going to be tricky to tell him our plans for the beasts.

After a month or so of rather idyllic existence, Milo and Donald ceased being lambs in temperament and look and became small sheep. As they further explored the acre, I was beginning to get anxious about my vegetables, which were so far safely hidden behind the house at the top end of the field. But it was only a matter of time before they found them, and then it would be only a matter of minutes before they snorted the lot. I had seen them scythe effortlessly through thistle, fern and nettle – picturing what they could do to sweet corn, beans and lettuce was making me sweat. I began to consider moving them back over the road to the main flock – after all, they no longer needed hand-feeding and besides, they would have further room to roam and would also be with their own kind. But the moving order wasn't activated until the moment they began to threaten the Beloved's fruit. Once they started chewing the lower branches of her apple and cherry trees that was it. They were off the premises.

We still saw them occasionally over the next few weeks. When walking through the fields, they would recognise us and come running in anticipation of extra food. The boys particularly appreciated this – they understood that air of being ever hopeful of a treat. But gradually the sheep reintegrated with the greater community and more and more often they stayed with the flock rather than come to us. Consequently they diminished a little in the boys' minds, but not in mine. The 'job' was looming.

The Farmer Friends had told me that male lambs at a certain age started to produce testosterone. Well, naturally they would, I thought, I can relate to that. They also told me that this gave their meat a strong, gamey flavour which could only be prevented by castration. I couldn't really relate to that, but I felt obscurely that I should be the one to do it. It was part of the whole process of raising, killing and eating an animal, and I felt that I would be shirking something if I didn't do it myself or at least watch it being done. I reconsidered that last option, imagining how my uncertain reputation in the village would be affected by gossip to the effect that 'yer man likes watching rams' balls being cut off'. No. It had to be me.

As it happened, the night before I got the shout, some friends from the city had been to stay with us, along with their children – one of whom, coincidentally, is called Milo – and we had stayed up late drinking unnecessary quantities of wine and beer. So I walked a little uncertainly up to the cross. As I approached the farm I became more and more anxious that I would be expected to tear off bloody gobbets of flesh and toss them over my shoulder. I steadied myself with the memory of a method Female Farmer Friend had told me about involving a type of elastic band placed around the scrotal sac – which then just withered away. This was what I hoped for, but it was not to be. It is a procedure only used when lambs are a few weeks old and is a way to prevent testes developing, not a method of detaching them. Donald and Milo were now too old for this and besides, I may have felt a little weird doing it – if someone tried to slip an elastic band around my testicles, I would probably consider it a sexual aid and thank them.

When I got to the farmyard, Female Farmer Friend handed me a huge pair of cast-iron pincers called a Burdizzo. I blanched briefly and my guts pitched like a small boat on a large wave. I wondered for a moment whether I was going to hang on to my breakfast. Male Farmer Friend popped Milo back on his haunches and held his front legs up to allow me access. I eyed the large scrotum nervously and, it has to be said, with a touch of envy, and composed myself. As it transpired, there was no need for squeamishness because there was no vomit or blood, and no apparent suffering on the part of the rams (or indeed, in the parts of the ram). The fierce looking jaws of a Burdizzo are very precisely calibrated. They do not shut completely and are designed to merely pinch the seminal tubes closed. They don't separate anything from anything else. I was much relieved and Milo and Donald seemed totally unconcerned. There was a very slight crunching noise as the tubes were nipped shut but not a bleat from the lads. They really are extraordinary creatures, I thought. Had they no notion at all of what had been done to them?

I walked backed down the hill with a firmer step than on the way up. I strode over to where our friends' big camper van was parked – we have a small house and they have a large family. They were surfacing blearily. Their kids had already risen and joined ours in running around the field. I couldn't resist chirruping, 'Good morning! I just castrated Milo!' I thought I was being hilarious. I explained myself and they were wanly amused. 'Ah … country humour,' they implied with their shared glance. 'How very droll.' My sensibilities, and thus my sense of humour, have changed over these ten years – coarsened, perhaps. I see the countryside more and more for what it is, in this part of the county at any rate – one vast anteroom to an abattoir.

Eating Lambs

We ate Milo's front legs the other evening and they were delicious. We slow cooked the shanks in shallots, wine and rosemary and had them with the last of the early potatoes and some fine beans from the garden. It was a simple enough meal, really, not being too complicated to prepare, but deeply nourishing and steadying.

Milo achieved the required weight and was killed on the eve of the feast of Saint Michael, or Michaelmas as it was once called. I thought this was amazingly serendipitous because it chimed with an old Michaelmas custom I have read about, but of course it is coincidental only because the seasons have had the same rhythm for centuries – despite what we are doing to change that – and things happen simply synchronously simply because it is the right time of the agricultural year. In the fifth century there was a king of Ireland called Leary whose son Lewy became mortally sick and died. This was in the time of Saint Patrick, who was busy trying to convert the country to Christianity. Like most missionaries of the early church, he was keen on 'magisterial' conversion, which was basically to demonstrate amazing prowess in magic, otherwise known as miracles, to kings and queens who would be so astonished that they would order the wholesale conversion of their people on the spot. The death of the king's son was Patrick's big opportunity. He called on the intercession of Saint Michael – because, being a totally amazing saint himself, Patrick could whistle up an archangel when he needed one. Michael duly brought the boy back to life. Leary was thrilled and we all became Christians. His wife, Aongus, was so grateful for her son's life that she wished to honour the archangel. So she vowed from then on to slaughter a sheep annually from each of her flocks and give them to the poor. It was ordained that all Christian converts should do the same on the saint's feast day. These became known as St Michael's Sheep or Cuid Mhichil in the older language. Well, we killed Milo, but we kept him ourselves and stuck him in our own freezer. However, we *are* heavily in debt so I think that counts.

I wore the coat when Milo and some other sheep reached the required weight of forty kilos and we brought them to the factory. It seemed the right garment and I felt that it could do with some sheep excrement to balance its strong perfume of Eau de Vache. I was expecting to be driving protesting creatures to their doom and therefore thought I should be prepared, both morally and sartorially, to be shat upon. It was not like that at all. The simile 'like lambs to slaughter' has its basis in reality. It was all very calm, efficient and stress free.

We arrived at the factory early in the morning with our batch. It is not an attractive building. I suppose in deference to public sensibility, abattoirs are mostly hidden away obscurely somewhere. Consequently their architecture is purely functional and eschews any notion of aesthetics. Up a tree-lined lane, largely hidden from the main road, we came to a large, concrete slablike building with a chimney at one end. This place apparently processed two and a half thousand carcasses a day. We unloaded the sheep at one end of the building and brought them into the first of a series of holding pens, which connected down the length of a long high-ceilinged room, at the other end of which was a curtained-off area with a ramp leading up to it. The air was warm and humid, brightly lit and full of the fug of sheep. Despite this I felt a slight inner chill – not faintheartedness exactly, but an awareness of the grave business in hand. Occasionally, silent, intent men wearing the green cotton suits you see in operating theatres emerged from the curtained area. We were moving our batch easily from pen to pen as we approached what I now guessed was the killing zone. A particularly large silent man in surgical greens suddenly appeared. He had a hairnet and a grim look. A phrase of my grandfather's came to mind – 'a face like a plateful of mortal sins'. I was startled when he spoke to me, staring at my coat: 'That'd stop the weather!' He nodded admiringly. I was very pleased to have the coat acknowledged. 'It'd stop radiation!' I answered. He grinned and disappeared back behind the curtain.

All the while the sheep had been calm and quiet – just the odd chatty bleat. Even now as we moved them into the last pen with the ramp, and you could catch a glimpse between long door-like plastic flaps of a splash or two of blood and upended, twitching bodies floating out of sight on moving hooks, the sheep remained oblivious. They went up the ramp, one by one, and silently disappeared behind the plastic flaps. From my slightly higher vantage point, I could see three or four of the green-suited men about their work, chatting humorously now – no more taciturnity or stern looks. I heard snatches of jocularity and the odd burst of laughter as their bright knives flashed and the surprisingly bright blood spurted. Over their shoulders I saw the steadily moving single file of dying sheep, suspended from hooks by one leg as the other leg kicked pointlessly. And not a murmur from those who were next – they trotted almost eagerly up the ramp. I was not completely surprised by this. I have been told of a sheep with a scab, which is tasty to a crow or magpie, being pecked to the brink of death whilst the sheep right next to it munches its lunch unperturbed. They are an odd species. They seem to like being together – they are flock animals after all – yet they are totally unmoved by each other's pain or pleasure. They are not unlike reality TV stars in that respect.

One of the green-suited men approached me – Milo was next. He had been earmarked, literally, to be separated from the rest, as I would be claiming him later. The man wanted to know did I want him stunned or not? I was a little confused, so he explained. The current batch was being killed according to the Islamic halal rite. The processes were pretty much identical – usually each animal was electrically stunned, hung on a hook, had their throat slit and then bled out 'in jig time!' Except in the halal fashion there was no stunning and the actual throat slitter in this instance was a religiously trained butcher who murmured the appropriate prayers as he wielded the knife. I glanced again at the green-suited men and now noticed that the one who seemed to be telling the best jokes didn't look especially Irish in the usual sense – he had a bright smile full of really good teeth and a burnished, healthy, fit demeanour. I now understood the twitching I had noticed earlier – a stunned sheep would have been hanging limp and comatose as it was conveyed out of sight. By this point I had anaesthetised myself somewhat and was about to say that I really didn't mind either way. Then I imagined telling the boys, if they ever wanted to know, what Milo's last moments were like. 'Stun him,' I said, and Milo wandered up the ramp.

Male Farmer Friend and I, having seen the last of our batch up the ramp, went around to 'the line' where the carcasses were being transformed from something that should be bleating in a field into something that should be hanging in a butcher's. We donned hairnets and disposable plastic aprons and rinsed our wellies in a disinfectant footbath. We then entered a corridor the length of a swimming pool, through which the line of suspended animals was slowly proceeding. At stations along the way, groups of men – it was an all-male workforce as far as I could tell – were busy at their particular tasks. A disassembly line, I suppose you could call it. The lambs were skinned, decapitated, de-hoofed and eviscerated in a swift surgical sequence. Then they arrived to be weighed at where we were standing. Milo was twenty kilos exactly. He then went to a chill room to hang for a few days before I could collect him to be butchered. Had I been a hardened Scot, I might have asked for his 'pluck'. This consists of the heart, lungs and liver, which are traditionally minced, then added to oatmeal and stuffed inside the stomach to make haggis. As it was, I only took the liver. We ate it that night with some broad beans and a nice Côtes du Rhône.

SAMHAIN

Falling Leaves

Seeing Ghosts

The first of November or Samhain was the old Celtic day of the dead. It used to mark the end of the farmer's yearly cycle and the beginning of winter, although, like all the seasons so far this year, winter seems a little slow in coming. When I opened our bedroom shutters this morning the ornamental cherry tree visible from our window was a blaze of copper and orange. On the horizon, Sliabh Buí was iron blue, but elsewhere the copses and patches of forest strewn across fields and hills displayed the full spectrum of autumnal hues. The forest of conifers at the top of our hill consists mostly of pine, but here and there it incorporates stands of larch. These are the only deciduous coniferous tree – they also turn and shed their leaves in the fall – so even the evergreen swathe has its patches of brown and umber. Brown is often a rather bland, boring shade, but when you see a variety of trees, bushes and shrubs clumped together and each one of them is wearing its own distinctive tone, the eye is dazzled by the complex, variegated palette of that single colour. Add touches of yellow, gold and red, and then illuminate the whole with the gentle light of the low sun, and the total effect stops the breath. 'The trees are in their autumn beauty' as Yeats wrote. At other seasons of the year, the forests and woods that surround us have an almost homogenous personality – they seem to acquire a common character and mood. The massed beech trees are generally cooling in summer and the ranks of pines are frostily silent in winter. On a particularly hot day the oaks along the riverbank exude a collectively drowsy air. Only in the autumn can I look at woodlands and see a host of individual trees.

There is a huge single beech tree just across the road from us. It has a still, stark beauty at twilight when its leafless branches form a skeletal silhouette against the setting sun. It seems entirely natural to me that seeing such sights in dwindling light would have suggested decay and death in earlier minds. This is the time when it was thought that the veil between this life and the next was at its thinnest. The Sidhe, or fairy folk, were abroad and on the eve of Samhain – now known as Halloween – the dead could be seen walking the earth again. Until quite recent times doorways would remain unlatched and food would be left out for these visitors from another world. Places would be set at the table for the family dead to join the evening meal and candles lit in the windows of rooms where loved ones had died. On the night of Samhain itself, or All Souls' Eve as the church calls it, it was usual to leave candles burning on family graves. This custom still survives in parts of the country, as I was alarmed to discover one crisp November night some years ago. Driving home from the city after dark, I stopped for moment to relieve myself against an old stone wall set back from the road. I noticed that it happened to be the wall of a cemetery, and when I glanced over, I saw flickering blue lights glinting on the headstones. I stopped in mid-flow, as did my heart for a cold, ghastly second, until I realised I was looking at votive candles in little blue plastic containers. I laughed aloud in relief – just to assure whoever (or whatever) might be listening that I was not easily spooked – and hurried home.

I remember reading a story once, I forget where, of a country cottage gathering on this night. There was a wonderful image of the souls of the family dead and departed perched about the room – on the mantelpiece, in the window nooks, even crammed onto the shelves of the kitchen dresser. It is a slightly unnerving idea. It evokes a picture of assorted aunts and uncles and the odd cousin (or 'odd' cousin) all eavesdropping disapprovingly. But it also suggests a sense of interconnectedness – an awareness of the dead amongst the living. In Ireland generally, and in rural Ireland particularly, there seems to be a readier acceptance of death than I remember experiencing in England. I don't think I am romanticising this, and I certainly don't ascribe any moral superiority to the differing ways of acknowledging death in these islands. But differences (very broadly speaking) there undoubtedly are. I am talking of death in the general community – when someone close to you dies, you deal with it personally and feel it uniquely: it is an individually felt grief that is nonetheless universally understood. But the recognised social formalities and received protocols of death are culturally variable. In England, whenever I heard that someone had died, I was always aware that perhaps 'one should not intrude'. This is no doubt derived from a very laudable wish to respect people's privacy at a difficult time and from an equally strong desire to avoid upsetting someone with gauche chat about the recently deceased. In Ireland there are no such scruples. If someone dies, you just pitch up regardless.

A few years after we arrived here, the neighbour up the hill who grew Christmas trees died unexpectedly. I didn't know him terribly well but had liked him as a gregarious, hearty man. He and his wife had called in with a few other neighbours the previous New Year's Day and he had joked and chortled as he drank numerous glasses of whiskey with admirable relish. He also got me out of a hole once, quite literally. I had reversed into the ditch opposite our house trying to negotiate the very tight turn in to our gateway (I came to driving quite late in life). He was passing at the time and, ignoring my embarrassment, nipped home and swiftly reappeared with a shovel. I watched him with double appreciation – of the favour he was doing me and of the deftness and vigour with which he handled a spade: it would have been impressive in a man with half his sixty-something years. He quickly dug around the back wheel until I had traction again. Then, waving away my thanks, he shouldered his shovel and walked back home.

The Beloved and I were both very sorry to hear of his sudden death. In an earlier life, I would have felt equally saddened by the news of such a man's passing and would certainly have attended his funeral. But I would have slipped in at the back, discreetly, feeling like an interloper, and maybe I would have written an awkward note tentatively expressing my regrets to his widow. I would never have presumed to call in to the house uninvited. On this occasion, the only point at issue was who should call in first. The Beloved decided to bake some banana bread to bring later so I went ahead.

Someone was leaving as I walked up to the house and the bereaved woman was still standing in the doorway as I approached. She nodded and smiled at me, then accepted my hug of condolence and showed me in. We sat at the table where a few people were already placed. One was the priest from the appropriate dispensation; the others were vaguely familiar. A companionable silence settled easily. When someone felt like it, they offered a reminiscence of the dead man: a humorous one, which would raise a laugh; or a characteristic one, which would elicit nods. The conversation rose and fell naturally. Occasionally the widow would weep. No one tried to stop her, someone might offer a hankie or squeeze her arm, and the tears would, in due course, subside. After an instinctually appropriate pause of acknowledgement, the quiet talk would resume. In the background, two women I recognised from the village had materialised and were silently making tea or coffee as required or offering sandwiches and cake to people as they arrived. This continued all day. People came and went steadily, sometimes bringing food and never staying too long. The evening before the burial in the village churchyard, they brought his body home. All the lights on the main street were turned off as the villagers lined the pavement and watched the hearse drive slowly by. Everyone came to his funeral.

As I write, the bereaved woman concerned is now herself in hospital. She is recovering well but a few days ago things were not looking good. I was walking the boys up the hill, and as we passed her house I told them that if we were to bump into either of her daughters we should ask after their mother. Older Boy said, 'If she dies, who will get the land?' He has yet to lose someone close to him and his notion of human death is still a little vague. The important question in his mind seems to be who gets what. This pertains as much to his parents' demise as anyone else's, as he once casually let me know.

My father has lived in Africa since I was a boy and I have visited him regularly over the years. On one trip, he and I were joining a small group to go on a wilderness trail into the bush – on foot and sleeping rough. We signed a disclaimer before leaving civilisation behind. This was to take personal responsibility for our own safety and it ensured nobody would be sued if we got eaten. We then spent four amazing days tracking rhino in the company of a one-eyed Zulu guide. Knowing my penchant for being dressed appropriately, dad bought me a terrific bush hat.

A wide-brimmed green-felt thing with a strip of striped antelope skin around the crown, it was perfect. During subsequent trips over the years it has been adorned with things I have picked up along the way – some snake skin, a vulture feather and the chrysalis of a giant African moth.

One rainy autumn afternoon when Older Boy was five or six, I was telling him tales of the African bush and I was pleased to discover they needed no embellishment because they were all true. I told him of a lioness with two young cubs lying in the shade of an acacia tree and how they slouched away as we approached; a flock of woolly throated storks flying lazily overhead as I sat on the banks of the Umfolozi River; and a zebra's pelvis stuck up a tree – the remains of a leopard's larder. He drank it all in. He laughed with delight at the story of a marauding baboon trying to steal my underpants on a beach on the Indian Ocean and how I only just won the ensuing tug of war. Later he asked if he could play with my African hat, and remembering how he and his brother had trashed a hat or two of mine already, I said that I'd prefer it if he didn't. I was expecting protest and pleading but he just shrugged and said, 'Ah well, I'll get it when you're dead,' and wandered off.

He and his brother bring this air of cheerful acquisitiveness to Halloween. They both grow very excited as the festival draws near, partly in anticipation of the night itself with its opportunities for amassing bucketloads of sweets – 'licensed begging' the Beloved calls it – and partly because in our house all mention of Christmas is banned until Halloween is over. It can never come too soon because thereafter it is open season for talk of Santa. This year they had decided quite early what look they were each going for and, with parties at school and at various friends' houses, they had the chance to refine their costumes to perfection. I got down my make-up box, which I have sadly not used professionally since May, and set to. Older Boy was an axe-wielding werewolf and Younger Boy was a vampire with a scythe. I downloaded pictures from the web and was pleased to see them both model themselves on the classics. Lon Chaney in *The Wolf Man* was one template and Bela Lugosi's Dracula, with particular reference to 'the big eyebrows, Daddy' was the other. There is pressure on the adults to join in and wear a costume to Halloween parties. Like most actors, the Beloved and I dislike fancy dress, but this year we surrendered and made a symbolic effort. I gave myself a livid scar across my face, which I felt was subtly macabre and just a little bit handsome in a virile, threatening way. She went for some devil's horns and a shedload of mascara, which I found strangely sexy.

Wandering the countryside in disguise on this night is a very old custom. It was believed that amongst the general throng of mischievous 'little people' abroad after nightfall the pooka was the most sullen, dark and dangerous. He was a kind of bogeyman and if he liked the look of you he might abduct you and whisk you away to fairy land. A scary disguise was seen as protection. He also apparently liked to urinate on briar fruits, so these were best avoided after the thirty-first of October. I suspect this story was a ploy to prevent children from eating rotten, worm-infested blackberries past their best. Although it would not necessarily deter Younger Boy, as he has been on intimate terms with many worms. If he finds a really big one, when I am digging potatoes for example, he likes to balance it on his curled upper lip and wear it as a moustache.

You don't have to live in the country for the dying of the year to bring to mind our common mortality. When I lived in

London I was out walking with a friend one blustery Halloween afternoon. We passed an old Victorian municipal cemetery and a tacit impulse made us both turn up its gloomy, cypress-lined drive. My friend had buried his dad earlier that year and I regret to say that I had missed the funeral. We wandered around in the fading light surrounded by stone angels, lichen-encrusted tombstones and mouldering family vaults. Neither of us had any relatives buried in that sombre, melancholy place, but each of us called to mind our own dead.

In earlier times, when minds were less cluttered by the tsunami of stimuli we are subjected to now, I can imagine how the desire to remember the dead could quite easily have been transmuted into a vivid sense of their actual presence. Bringing something to mind can often bring it to life, and an individual's notion is often affirmed by a communal acknowledgement when gossip turns idle talk or rumour into fact. So it can be a very short step from wishing to see a lost loved one again to imagining the countryside infested with ghosts, ghouls and joking sprites. With all the supernatural trickery and anonymous, magical mayhem of the night, it was inevitable that people wanted to join in. Practical jokes came to be expected, if not welcomed, and 'trick

or treating' evolved. If spirits were abroad, people dressed like them and moaned at cottage windows. Hollowed-out turnips had fierce faces cut in them and, lit from within by a candle, they were placed in a spot strategically chosen for its high scariness factor. These jack-o'-lanterns ended up in America as part of the cultural luggage brought by generations of Irish emigrants and – as so often happens in that capacious country, where imports are absorbed, increased in size and then exported again – the diffident turnip was exaggerated into an ostentatious pumpkin and brought back to these islands in Hollywood slasher movies of the seventies.

Halloween was a night of licensed misrule – a brief riot of semi-institutionalised subversion – but very little trickery of any sort is perpetrated any more. The kids just gallop from door to door, bellow a short rhyme and grab as much loot as they can. I admit that these were also my key concerns when I was a child but I also remember a sense of mischief in the air and I certainly remember being scared. We were not usually out and about after dark and that in itself was exciting, especially as there were no adults with us. We roamed around our neighbourhood in gangs of feral midgets. The big thing was to scare and jeer the other gangs as we passed them. There were often tears and it was sometimes a relief to get home and crack a lot of nuts and eat a lot of apples. These days the boys come home with very few nuts and no apples. This suits them perfectly, as they have little interest in accepting such traditional offerings – although they will have a go at apple bobbing and such like. Perhaps these seasonal games developed as an early attempt at getting children to ingest a bit of vitamin C. Or perhaps children were just more appreciative of fruit then, with few other sources of sweetness. It must have been hard to get them their 'five a day' in those days – especially if the phantom pissing pooka was ruining all the hedgerow fruits. Our late-producing apple trees are groaning with excess weight at the moment, but we cannot induce the boys to eat the fruit unless they are spinning on a bit of twine or bobbing in a bowl of water.

There doesn't appear to be as much interest in telling scary stories either. Which maybe is a good thing, as a thrilling tale well told can lead to nightmares in the boys' room and recriminations in ours. The closest I came this year was telling the story of *Macbeth* at the boys' school. I have been in two productions of the play, though regrettably not as the lead, and had once told the boys the thoroughly ripping yarn. As it happens, the schoolchildren had been looking at the witches' charm from the play as Halloween approached, and Older Boy mentioned to his teacher that his dad did a great version of the story. So after school one day I was asked if I would tell the rest of the children the tale some time when it suited. Of course I said yes – this might be the only chance I was going to get to 'give' my Macbeth – and we set a date.

This was not my classroom debut, as I have appeared at the school before: playing Odysseus, Achilles and the goddess Athena – I made a big impression in this last role. Older Boy got me that gig too. He should definitely be my agent. He had been talking a lot about Holy God at the time and I thought he should know about alternative concepts of divinity. You can't beat the Greek myths for gods galore so I told him about Olympus and Troy and Agamemnon dead. When they subsequently read about Theseus and the Minotaur at school he pronounced the version 'really boring compared to my Dad's.' He should be my publicist too. So I was asked down to the school to give them my version of the story of the man with a bull's head lost in the Labyrinth. I threw in the Iliad as a curtain raiser. The reviews were excellent – the boys and girls loved it and my son was very proud.

Feeling consequently that I knew the crowd when it came to staging *Macbeth*, I thought that this time a bit of audience participation might be fun, so I suggested to the teacher that they might learn the 'Double, double, toil and trouble' bit and we could chant it together during the scene. We also thought the children were ready to be exposed to some of the great poetry in the play, so she agreed to familiarise them with Macbeth's famous soliloquy at the end, to help them grasp its sense if I spoke some of it. The thing went down terrifically and I enjoyed it as much as they seemed to. They squealed and groaned at the witches and the bleeding dagger. They shouted and roared at the murders and Banquo's ghost. They zealously chanted the incantation's 'wing of bat and tongue of dog' with gleeful disgust. They were hooked and I was storming it. The big speech was coming up and my intention was to speak it simply but truly and perhaps give them a glimmer of the beauty of the images and the pitiable desolation of Macbeth's mind. I started softly: 'Tomorrow and tomorrow –'and they were in like rats down a hole. They had learnt it by rote and they gave it all they had. They bellowed the lines and drowned me out entirely. Ah well … I was only mildly disappointed. Besides, I did the speech that night anyway – to myself in the bathroom mirror. I was marvellous.

My mother's second husband was from Yorkshire, and when I moved there at the age of nine, I missed Halloween. October was not quite over and the date hadn't passed, however I was extremely disheartened to discover that they didn't really 'do' Halloween in this new place. It had been upsetting enough to leave my extended Irish family, and saying goodbye to my father was achingly sad, but this was too much. I had been promised that everything in England was going to be great but I began to have my doubts. I wanted to go home. I was reassured that there was a much better event in store – Bonfire Night! With hot dogs, baked potatoes, toffee apples and fireworks! This sounded attractive, and when I heard that they would be burning a 'guy' on the bonfire, they had my serious attention. I was unfamiliar with Guy as a Christian name and had never heard of Guy Fawkes. To my ears the plan seemed to be to burn a bloke. What bloke? And what had he done? And was it even a specific bloke? The talk was of 'a' guy, not 'the' guy. Did they just grab someone from the crowd and fling him on the flames? They don't mess around in this country, I thought. They didn't seem to have any nuns but it might be equally scary. I had better go along with this and see what happened.

As it turned out, the only roasted carcass I saw that first Guy Fawkes Night was that of a whole ox being barbecued on a spit in the local park. I had some in a bun – a succulent slice of hot beef with fried onions – I can still taste it. We joined the crowd gathered around a huge public bonfire. The flickering orange glow warmed our faces as our backs shivered in the dark. It was with a mixture of relief and disappointment that I realised the smoke-wreathed figure perched at the top of the flames was just a scarecrow and not a bloke at all, but I couldn't really express any frustrated expectations because my jaws were gummed shut by slabs of Yorkshire toffee. And the fireworks were great. Bangers and sparklers were all that I had encountered before. I wondered why we hadn't celebrated the fifth of November in Ireland. It was only when I got older, and I learnt the history of the Gunpowder

Plot and its foiled attempt to assassinate a Protestant king, that I realised the symbolic burning of Catholics wouldn't have been a crowd pleaser back where I came from – on most of the island, at any rate.

When I started at university in the south of England, I went with a group of other first years to a nearby town for Bonfire Night. We had heard they really knew how to do it there. It was an evening of fierce enjoyment, beginning with many pints of fine local ale, followed by much leaping over subsidiary bonfires which were scattered about the place and finally, a torchlight procession through the town to the main pyre. The culmination of the night was the ritual burning of an effigy of the pope. Which I thought made the point a little emphatically. I'd like to burn an effigy of a particular casting director.

Up at the farmhouse a few nights ago, we had an impromptu seasonal supper of colcannon and sausages followed by a

barmbrack. These are traditionally Irish. We also had buckets of wine – which isn't. The colcannon was a heartening, steaming dish of buttery mashed potatoes mixed with cabbage and scallions – deeply fortifying on a cold night. Barmbrack, or *báirín breac* more properly – someone once said 'the Irish are great spellers but terrible pronouncers' – is a speckled fruit loaf. Correctly made, it also contains little tokens which supposedly foretell the future of whoever gets them in their slice.

A ring signifies a wedding and a button means you remain unmarried; a coin promises wealth and bit of rag indicates poverty; and depending on the part of the country, a little chip of wood means either you will work hard till the day you die, or you will spend your life being beaten by your partner. Hoping to avoid either destiny, I didn't want the last one and I was amused to get a button, which signifies that I will remain unmarried. The Beloved laughed when she got one too. We interpreted this as fate showing its approval of our continuing sinful cohabitation. We have never married and have no real wish to. I have found ways of expressing my love for her over the years, sometimes in poetry, and have never felt the need or desire to stage a 'romantic' proposal. Besides, I have a strong suspicion that if I dropped to one knee and offered her a ring, she would guffaw dismissively. Nonetheless, many years ago we reached a stage in our relationship when we both knew we wanted to live our lives together and marriage was a word that was hovering listlessly about. So I casually mentioned it one day. We are both children of 'broken homes' – to use that silly, reductive cliché – and both our families have complex histories with a freight train's weight of baggage. She pondered the word for a moment, looked at me and said, 'Imagine the top table at our wedding.' We both shuddered and dropped the subject. The boys raise the topic from time to time. Their current position is that neither of them wants a wedding because the Older Boy doesn't want to wear a suit and Younger Boy 'doesn't want to see any smooching' – although on reflection he declared that he might accept the idea if he could wear a kilt and carry a mace. But for the moment, the barmbrack has spoken.

Traditionally, there were many other things that 'spoke' at Samhain. Divination of some sort or other was widespread and various. My reading of signs in the night sky or the orchard is elementary and narrow compared to the inventiveness displayed by my forefathers. They could tell the future by which way a nut roasting in the fire jumped; or by the shape molten lead made as it dropped into cooling water; or by the pattern a snail's trail made on a plate. And if a young woman ate a salted herring in three bites on that night she was sure to have a dream in which her future husband would be conjured up offering a glass of water. As a small boy, I knew for a fact that if you said your name backwards into a mirror at midnight on Halloween, the devil would appear over your shoulder. I also remember that every year in early November the same 'news' would circulate of a country dance that somebody's relative had attended. At the height of the festivities a handsome dark stranger had appeared. He had laughed with all the men and danced with all the women. Everyone was having a fabulous time until someone noticed in the mad, jubilant crush that the charming outsider had cloven feet.

In a more pragmatic vein, forecasting the weather would have been useful in old rural communities, and there were many ways to do this. The direction of the wind at midnight on Samhain supposedly indicated the prevailing wind for the months to come, which seems to merely demonstrate the instinct of a good bookie. It was also possible to read the moon: a clear moon meant fair weather; a cloudy moon meant rain; and if the moon was obscured by rolling clouds, a storm was coming – which seems to me to be pretty bleedin' obvious. Male Farmer Friend has proved himself to be notably accurate in anticipating changing weather and I was wondering how he did it. Was it the behaviour of his cattle in the field or the colour of the berries on the holly or the rowan? I imagined a deep-grained, intuitive country sense of the seasons at play. Over a pint one night he told me his secret – he logged on to the website of an amateur meteorologist in New Zealand.

The autumn breeze is shifting as I write, and the sky is darkened by ragged, wind-blown rooks. Above the hill they buffet, swoop and veer, cawing cacophonously. When Older Boy was a toddler, he described this noise as 'crows laughing'. I find it a pleasing sound and it still reminds me of being a fresher at university. My first term in residence was spent living beneath the shelter of ancient elm trees. Standing in comparative isolation on the Sussex downs, they had survived the ravages of Dutch elm disease, which devastated the native English population of elms in the mid-twentieth century. I was very saddened to hear later that they had not withstood the great hurricane of 1987. When I knew them, each tree was a giant rookery. In my subconscious mind, the clamour of crows is associated with the start of the academic year, so that even now, as the earth seems to be withdrawing into itself, I have a vague sense of a new beginning. I must go and shoot something.

ELDERBERRY CORDIAL

Ingredients

elderberries
sugar
cinnamon sticks or ground cinnamon
Root ginger or ground ginger
cloves
booze – your choice

Method

Pick plenty of elderberry heads. The stalks are not good for you, so you need to get the berries off the stalks. The best way to do this is to hold the berry head by the central stalk and, with a dinner fork, comb the berries into a large bowl below. Cover the berries with water in a saucepan and simmer gently for 30 minutes – you can mash them a bit too, to help them surrender their goodness. Strain and measure.

For each 570ml of juice, add 225g sugar, 1 tsp of ground cinnamon or a few cinnamon sticks, 30g of chopped ginger or a teaspoon of ground ginger and 8 cloves.

Or just add anything else that reminds you of Christmas. If you don't like spiced drinks, just leave them out and enjoy the elderberries in their purity.

Simmer all the ingredients together for another 30 minutes. Add a good glug of brandy/cognac/whatever you prefer.

Strain out the whole spices, bottle in sterilised glass bottles and seal. Alternatively, store in small plastic bottles and freeze. Remember to leave space for expansion in the freezer. Once open, store in the fridge.

A couple of teaspoons of this a day will (should!) keep you free from colds and flus all winter.

Chasing Rabbits

I have had a shotgun licence for three years now. I had to go through a complex procedure in order to get it, which is as it should be. I was required to obtain character references from 'respectable' members of the community – no actors then – and I also needed permission to shoot on the land of local landowners. A cousin and the Beloved's sister, both with 'proper' jobs, supplied the former and Male Farmer Friend was happy to supply the latter. He was quite keen for me to get my eye in by shooting as many crows and magpies as I could. These are terrible pests to him. I can't bring myself to do this as I only feel happy shooting something we can eat, so I take speculative shots at rabbits or pigeons and go after pheasants when they're in season. The first two are classed as vermin, so you can shoot them year round. My introduction to shotguns was when Male Farmer Friend took me with him to get the pigeons which were ravaging his cornfields. I have fired an occasional blank round in my professional life, but the only time I had ever used live ammunition previously was on an Israeli army base.

In my twenties, I was cast in a movie being shot on location near Jerusalem and I thought I had made it. It was a contemporary Israeli story but the principal cast had all been flown in from London and New York, much to the understandable annoyance of local talent. The bit parts, or 'day players' as they are now more diplomatically termed, were all Israeli, and I now understand how they felt. Many big projects filmed in Ireland these days come with cast attached and all that is available for Dublin-based actors is a meagre scattering of small roles. Players living in far-flung villages often don't get a look in at all. But I made my choices and am not too bitter. Back then I was oblivious to any resentment. I was excited to be working and, as I was living in Brixton at the time, it was a pretty exotic location. The director of the film felt that since we were all playing Israelis we should have a taste of Israeli life, and as military service is compulsory there, a day on an army base was prescribed for all foreigners. We were met by the officer commanding – an intimidating, much-decorated colonel. He gave us lunch and told us a story.

A scorpion and a fox are at the side of a flooding river, both eager to cross. Because he can't swim, the scorpion asks for a lift on the fox's back. The fox objects that it is in the scorpion's nature to sting and he doesn't want to risk it. The scorpion points out that, in this case, it is not in his interest to sting because if the fox drowns, so does he. The fox agrees, the scorpion hops on and halfway across, he stings the fox anyway. 'Why?' asks the drowning fox of the drowning scorpion, who shrugs, grins and says, 'This is the Middle East!' The colonel thumped the table laughed maniacally. Later that afternoon on the assault course, he thrust a Galil semi-automatic rifle at me and invited me to take a shot at a life-sized cardboard cut-out of an armed soldier – the type of thing you see in *Rambo* movies. I asked him if it was a scorpion or a fox. He grinned and said, 'Does it really matter?' I admit to a slight thrill when I hit the target, and the colonel cackled.

I had the same beginner's luck when I went after pigeons with Male Farmer Friend. We set ourselves up in the field opposite our house, close by the hedge and behind some camouflage netting. Pigeons have very keen sight – a slight twitch of something unfamiliar and they're away. We also spread some very realistic decoys amongst the stubble about thirty yards from our hide to encourage passing pigeons to land. Not that they needed enticing – I had seen flocks of them gorging on grain over the previous few days. Male Farmer Friend showed me the rudiments of firing a shotgun: how to hold it properly; how to flip the safety catch just before firing; how to eject a spent cartridge. Then he potted four in quick succession and passed the gun to me – stock first, of course. The essential safety rule is 'Never, ever let your gun, pointed be, at anyone! All the pheasants ever bred, won't repay for one man dead.'

I loaded both barrels, waited for some pigeons to land, took aim and fired. I missed, naturally, but wasn't conscious of any disappointment because all I could think was 'Jesus Christ! That was really, really loud!' I inserted the earplugs I had brought but had forgotten to put in and relaxed a little. We didn't stay out much longer, as it was cold and we were keeping as still as possible so blood circulation was slow despite the undeniable jolts of adrenalin. I only fired four or five more rounds but I had two hits before we left the field. I winged the first one. It cartwheeled to the ground and fluttered for a moment in a flurry of feathers until Male Farmer Friend dispatched it with a blow of a small cosh called a 'priest'.

My second kill, because that is what it was, had wheeled in to land and then suddenly soared away fast, high and towards our hide. I followed it with the barrel of the gun and until it was almost overhead, then I pulled the trigger. It dropped like a stone and hit the ground with a surprising thump just a few feet away. I felt a short, sharp atavistic thrill but no guilt. I had often eaten game birds before. Someone had shot them and there is an undeniable satisfaction in doing it yourself.

The Beloved and I had pigeon for dinner that night. I filleted two – a brace – then we pan-fried the breasts and ate them with a mushroom and Marsala sauce and our own vegetables. It was delicious and as fresh a meal as is possible, which is, to a purist, strictly speaking wrong. Game should be hung for a few days – for practical reasons. Wild fowl have much tougher muscles than their more sedentary domestic cousins and hanging tenderises the meat. It also develops and intensifies the taste. The gamey flavour that hardcore gourmands seek in very well-hung flesh is essentially the whiff of decomposition. We are still novices in this regard, and I hung the remaining pigeons in the garage for a mere four days before plucking, gutting and freezing them. They ended up in a richly unctuous pie.

The first time I shot a rabbit was much less successful in terms of the ratio of hits to misses but it was a huge amount of fun in a hillbilly kind of way. It was a cloudy November night and I was well wrapped up against the cold when Male Farmer Friend called for me in his four-wheel-drive pickup truck and we headed up the hill. We turned into the hill field and drove around the inside of the hedge with headlights on full. Then we spotted the small, dark shapes of a few rabbits feeding in the grass. Male Farmer Friend abruptly stopped the truck, holding the rabbits in the beam as they peered curiously in our direction. I noticed the ruby glint of their eyes and had hardly formulated in my mind the phrase 'a rabbit caught in headlights' before Male Farmer Friend had flipped open his door, whipped up his rifle and shot two. The rest of the rabbits hopped into the hedge.

This was pretty impressive shooting and neither of us expected me to match it when he handed me the gun and said, 'Your turn.' I got up into the bed of the truck and stood behind the cab, facing front and holding tightly with my left hand. In my right I had the rifle, pointing skywards with the butt snug against my hip. Then we drove off, briskly, bouncing across the furrowed field. I had to restrain myself from shouting, 'Yeehaa!' We skidded suddenly to a halt with the headlights picking out a few more rabbits. I quickly braced myself, leaned forward on the roof of the cab and aimed at the nearest and biggest of them. The truck had excellent suspension and the cab was still rocking slightly from our sudden stop, so I inevitably missed with my first shot. I fired again. The rabbit was still unharmed, but now it turned and hopped a couple of paces then turned back to look at the strange light once more. I fired and missed again … and then again. It hopped closer to the hedge and stopped one last time to glance back. I fired a final, pointless shot before the rabbit disappeared. The cab of the truck was still shaking. I realised this was because Male Farmer Friend was quaking with suppressed laughter. 'That rabbit's off to buy a lottery ticket!' he chortled as he started up the engine and we swooped away. I clung on and the breeze whistled in my ears as we rocked around from field to field, pausing periodically for me to fire futile bullets into the empty meadow. It was a completely exhilarating, boyish romp in the dark and I eventually managed to shoot two rabbits before we ran out of ammunition.

The habit of shooting animals by bright light after dark is known as 'lamping' and is frowned upon in sophisticated shooting circles, but to farmers it is an efficient way of controlling pests like rabbits or foxes – although not with me shooting, of course. However, I have to say, it is not as easy as is commonly thought. The popular notion is that rabbits, in particular, are so hypnotised by the unexpected light that they sit there totally stunned whilst you pick them off at your leisure. It is not quite like that. They are certainly perplexed by the unfamiliar dazzle and blaze, but after a few curious seconds they hop away. You have to be able to take your shot quickly and my shooting wasn't really up to it. I needed some practice.

Later that week, I took the rifle out in daylight and lay in wait in the lime-kiln field, where Male Farmer Friend told me I might be lucky. I chose my spot carefully. I was well hidden in the shade of some tall ferns and I had vantage over a long bank, along which the rabbits emerged to feed in the late afternoon. The important thing about the bank was that if I missed, which I acknowledged as a possibility, the bullet would drill into the earth. Bullets can travel a long way and it is vital to be aware of that when shooting. From the back of the pickup I was always shooting safely downwards into the ground – you should never fire a rifle without being confident your bullet is harmless if it goes off target.

Some time passed before I spotted the distinctive ears of a rabbit sticking up amongst the clumps of gorse perhaps forty yards away. I got it in my sights and pulled the trigger. A little puff of dust spurted a few inches to the right. The rabbit's nose twitched inquisitively as it looked at where the bullet had hit. I adjusted my aim and this time gently squeezed the trigger. Now there was a puff of dust to the left. The rabbit stopped chewing and stared quizzically at this new spot. I remembered to breathe and fired again. This time my line was good but the shot fell short and I saw a pebble ping upwards just in front of the now thoroughly puzzled rabbit, who turned its back and loped away. I saw its white bobtail gleam in the telescopic sight as I fired one last time. I got it. I shot it up the hole. If it had happened to yawn as I pulled the trigger, the bullet might have passed clean through it.

I walked stealthily over to ensure it was dead. You have to be wary approaching a shot rabbit. Seriously – its dying kick can apparently break your wrist. You should approach from behind so that it can't see to lash out. Feeling only a little foolish, I snuck up on it. It was clearly deceased so I decided to paunch it there in the field. This is the appropriate term for gutting a rabbit. I have learnt that there are specific words for this process depending on what animal is being eviscerated at the time. You 'clean' a fish, 'draw' a chicken or pheasant, 'paunch' a rabbit and 'gralloch' a deer. I had been told that a rabbit's innards started to rot quickly, and that if it was going to be hung for any amount of time at all, the meat would be tainted with a flavour well beyond even an extremist's definition of gamey.

Feeling a strange need to appear to be doing a thing properly, even if it was only in my own eyes, I decided to act promptly. I suppressed a pang of squeamishness and picked up the dead rabbit by the ears. I applied the sharp tip of my hunting knife to just below the sternum and, in one firm movement downwards, slit the belly open. The entrails flopped out in a steaming pile onto the grass. They were still attached, of course. There was nothing for it but to put my hand into the body cavity, grip the slippery, warm mess tightly and yank hard. Out they came and slithered to the ground. There was a pungent, sweet, rich stink of blood, half-digested grass and rabbit shit. I wiped my hand as clean as I could on some dock leaves, picked up the rabbit, shouldered the rifle and walked home, not sure whether I had the stomach for any more that day. However, my appetite was unaffected. It was a young and tender rabbit and after we had roasted it with potatoes, lemon, shallots and rosemary, I ate it gratefully. I have since shot and drawn or paunched many pheasants, pigeons and rabbits, but the memory of that early, oddly intimate kill is still vivid.

Not every rabbit I have subsequently shot has been as delicious. They can sometimes be leathery and dry, depending on the time of the year and the size of the rabbit. The older and bigger they are, the tougher their meat. And in the summer months, when they are rampantly breeding, the hormones surging through their bodies give the meat a not-so-pleasant flavour. Now is the best time. Male Farmer Friend will often leave a few at our door, knock and drive off. If the Beloved answers the door and I hear a squawk, I know there is something for me to skin and joint, and possibly some blood on the step for me to scrub. Our doorstep is often smeared, to her audible distaste, with substances she doesn't want in the house, but she particularly dislikes it bloody.

The eleventh of November is the feast of Saint Martin or Martinmas. He is the saint who supposedly conferred the monk's tonsure on Saint Patrick. In memory of his holy haircut, Patrick ordained that every monk and nun in the country would be given a pig to be killed and shared with their local community on Saint Martin's eve. This tradition spread and evolved over time and eventually, according to their wealth, people killed a cow, or a sheep or a goose or even a mere chicken, and shared it with the poor. Most of the country customs I have seen or heard of have a basic pre-Christian element which has subsequently been adopted then adapted by the church. The pagan aspect of Martinmas that I particularly relish is the ritual dropping of blood from the slaughtered animal at the threshold of the front door. This was a way of warding off ill luck. Shortly after I discovered this, I left some rabbits at the door prior to skinning them, and the Beloved complained loudly of the gory doormat. I counterclaimed protection from evil spirits. Her response discouraged me from adding the detail that on Martinmas you also sprinkled blood in the four corners of the house.

I don't have a dedicated space for dealing with game. If I have rabbits to skin, I go to the piggery, where I can hang the bodies on the open door and use the chopping block I usually split logs on to chop off their paws with a hatchet. You have to do this in order to pull the skin off smoothly. I can do it reasonably well now but my early attempts at skinning a rabbit were like wrestling a greased ferret. But you learn. I have also learnt not to feel guilty of littering when I throw the discarded heads, skins and paws into the next field. Male Farmer Friend suggested I do this when I asked him what to do with all the bits we weren't going to eat. I remember thinking that this didn't sound ecologically virtuous or particularly hygienic, but he assured me that foxes, badgers, crows, magpies and even rats would make short work of the lot. He was, of course, right.

Once I threw some rabbit leftovers across a hedge opposite us. A skin caught on a branch and hung there, slightly out of reach across the ditch. I made a mental note to come back in wellies and pluck it down later. Then I forgot. The next morning as we were walking the boys to school I saw the skin still hanging there, picked clean of flesh and blood. You could have worn it. This gave the Beloved an idea. At her request, I started saving rabbit skins. I cleaned them roughly and froze them until a dozen or so had accumulated. Her plan was to tan them and make a lovely, soft rabbit-skin bedspread for our room. I pointed out we only had enough to cover a foot stool. She swatted this aside, so I thawed what we had and she steeped the skins in a bucket of alum and salt solution where they mouldered for over a week. I was a more regular visitor to the garage where they were palely loitering than she was, and I occasionally asked how the tanning was going. She told me it was all in hand, until eventually, the bucket mysteriously disappeared. She later admitted that she had flensed one skin – that is, she scraped all the flesh off – which was 'very hard work', but she had nowhere to dry it, so she grew discouraged and ignored the rest until they were beyond use. She said she now understood the pain of her Apache sisters trying to tan buffalo hides in wet weather. She wistfully remembered a moment in the movie *Dances with Wolves* when Kevin Costner jubilantly jigs around a large wooden frame on which a bison pelt is drying – clearly delighted with the thing. Is this a hint for an interesting but useful Christmas present?

Dealing with pigeons and pheasants is much messier than rabbits. There is always a snowstorm of feathers whether I am plucking the birds or just filleting the breasts. I set myself up with plastic bowls and super sharp knives at an outdoor wooden table under a thriving cherry tree. The table and the tree were gifts from the Beloved's mother who feels proprietary about both. She beams with satisfaction when the table is surrounded by blossom, but she sniffs with disapproval when the table is surrounded by feathers. She has a point – the blossom rots gradually and invisibly into the grass; the feathers don't, and they get everywhere.

Plucking fowl efficiently is a skill I have yet to master. It is easiest if the bird is just dead, when the feathers come out quite easily. But this is not the case with game birds, as they have usually been hanging for some time. By this point the feathers seem to be stuck more tightly, which is strange, because the skin is more fragile and liable to tear. Your fingers get moistened by the exposed flesh; they become bloody and very sticky. Then they get clogged with small downy feathers and you can't grip the bigger ones properly – the wing and tail feathers feel really firmly rooted. Then you whimper with frustration. I now avoid plucking a whole bird if I can. I usually slice out the breasts and skinned legs of a pheasant and take only the breasts of a pigeon (there is very little meat to speak of on their legs.) Then I dispose of the bodies in the usual way by tossing them over the hedge. Although one large batch of pigeon carcasses ended up in the septic tank instead. We had just had it emptied and I threw them in. There was a good reason for doing this besides laziness. It used to be common to throw a dead sheep into a freshly cleaned septic tank in order to kick-start the required biological processes and I figured a small flock of dead pigeons would do the same job. It seems to have worked well as we have had no problems with it recently.

There have been problems with the septic tank in the past, which is really not what you want. Not being on the mains, it is how we deal with waste water from the kitchen sink and the bath, and, much more crucially, it is how we dispose of sewage.

It is a deep concrete-walled pit with a concrete lid set just below the level of the grass about twenty-five yards down the slope of the field from the house. It is fed by a pipe with a gentle gradient downwards, so that effluent flows consistently away. That is the theory – in practice it has backed up numerous times. The first time it happened, I was away working. I was as supportive as possible to the Beloved on the phone when she told me the problem, and deeply relieved that I wasn't there to deal with it. She called Male Farmer Friend and he pitched up with the appropriate rods and unblocked the pipe enough to allow the flow to resume, but he discovered a severe bottleneck a yard or so from the tank which he said needed dealing with properly. His guess was that the pipe had caved in and would only get worse with the passing months. The months duly passed and I conveniently forgot the need to 'properly' deal with it. When I called up to him to borrow the rods for another unblocking, I avoided answering when he asked, 'Have you not dealt with that septic tank yet?' I just shrugged sheepishly and thought of butterflies to banish the job from my mind.

I got away with this for a few years until the lavatory bowl started to back up and I had to act. My hope was that I would only have to dig down a foot or two to uncover a small section of collapsed pipe, which I could then easily cover in a swift patch job using a short length of piping I had found in the garage. I got the shovel and pick and started digging. It took most of a day of back-breaking toil to get down through stones and rocks and thick clay. I had to dig a trench six foot long, four foot wide and five foot deep to uncover the problem. The hole resembled a grave, which was apt because I would have liked to bury whoever had laid the pipe in it – preferably alive. There was no broken, caved-in bit anywhere – the pipe had been bent to fit and as a result it was pinched almost closed. I was surprised it had worked at all. It should have been laid at a slight but consistent downhill gradient so that the lower end was exactly level with the mouth of the septic tank. However, the pipe had reached this depth at about five feet from its terminus – from there it descended a further three feet until it had been forcibly twisted up again to connect with the tank. This meant that for the last section the flow was against gravity, against nature and against all principles of decent plumbing. I was sitting in the bottom of the hole, cursing and whinging, when Male Farmer Friend looked down at me. I hoped he would help me or say something like, 'Great work! Leave the rest now and the fairies will finish the job.' But he didn't. He just nodded grimly and said, 'Well, you're over the worst.'

On this rare occasion he was wrong. The bent pipe had to be replaced, but before I could get out of the hole, the Beloved was off to the hardware store to get the required length of new piping, thus robbing me of my last remaining delaying tactic. I began to saw off the useless bit. The residual sewage in the pipe washed around my feet and my whinging turned to whining. Thankfully I was wearing wellies and I did derive a degree of comfort from the fact that the turds I was eyeing were at least those of people I loved. I banished self-pity, resigned myself to the task and heaved the bent pipe free. The Beloved reappeared with the new pipe and an angled collar to attach it. With her help, this last leg of the job was quick and easy. She went up to the house, flushed the loo and nipped back eagerly. We watched the unimpeded flow with glad hearts. It was a beautiful thing. I filled in the hole and replaced the grass-covered sods I had cut for that purpose earlier in the day.

I was proud of myself. In my city days I baulked at using a plunger to unblock a toilet and now I had finished a filthy job to my complete satisfaction. I suppose my definition of dirt has changed – partly out of necessity and partly with a changing environment. I don't particularly mind if I step in cow shit in the country but I hate stepping in dog-shit in town. The dirt seems somehow cleaner down here – to me at any rate, though not everyone feels the same.

It seems muck and faecal matter are acceptable in a field but disgusting in an urban environment – unless they are flowing through a sewer, of course. And like our septic tank writ large, sewers also need maintenance from time to time, as I know from experience. I spent half of one summer vacation from university working for a company that specialised in 'de-scaling contracting' – which is industrial jargon for 'shit shovelling'. The work consisted of cleaning things on an industrial scale. I was part of a gang who shovelled sludge out of blast furnaces, coking plants and chemical works. The most memorable week was spent underground, hosing down Victorian sewerage tunnels. We used high-powered hoses as we proceeded along the walkways just above the river of effluent. The sewer rats fled from us as we made our way from manhole to manhole, through which we were detaching and reattaching the hoses to hydrants up at street level. Periodically, the rats' escape would be barred by brick walls across the walkway. We could negotiate these by means of the metal rungs placed around the edge for that purpose. The rats, of course, couldn't, and when we got to within a certain distance they would try to leap over to the walkway on the other side. This was when the fun began. Mostly we missed, but with practice you might hit a rat in mid-leap with the jet from the hose and it would be shot a considerable distance away down the tunnel. Years were to pass before I tried to shoot a rat again, and that was when a horde of them overran our hen house.

Feeding Chickens

I have finally accepted that I put the chicken run in the wrong place and it has to be moved. Currently it is out of commission and covered with weeds. It stands at the top end of the field under the lilacs and between the compost bins and the kitchen garden. The vegetable patch was promoted to 'kitchen garden' when it got a path. I paved it with some second-hand flagstones which I laid on a layer of supposedly weed-proof matting. It is partially effective – now I only have to mount a sporadic blitzkrieg against the weeds instead of the old trench warfare. The compost bins are mouldering a little, but they are a vast improvement on what had been there before. I used to inaccurately call this 'the compost heap', which is was what it was only in my mind – a platonic mound where garden clippings and potato peelings decomposed in a cleverly managed fashion into the nutrient-rich black gold lusted after by all gardeners. The mundane reality was a big pile where I chucked weeds so that they could grow to gargantuan proportions and mock me. This was also upgraded when I built two adjoining wooden compartments – each one a yard square and a yard deep and both covered with a bit of old carpet. One is left undisturbed to cook green matter down to good compost while the other is in daily use as the receptacle for garden waste.

Everything was organised and industrious and the agricultural quarter of our acre was starting to work well. But there was still a bit of wasted space near the hedge and we decided to get a few chickens to complete the picture. It seemed an obvious next step. Four years of baking bread, growing vegetables and making pickles had passed – we were ready for livestock. Always the catalyst, the Beloved announced that she wanted a hen house for her birthday. The location was obvious – the aspect would be excellent, facing east so that the morning sun would wake the hens early and get them laying; and the construction would cost next to nothing, as apparently I would be building it. I figured Quigley could knock up a chicken coop easily enough – all that was required was a box with a few holes and a roof, maybe a perch – chickens like to perch, don't they? No. This was inadequate. The Beloved asked me to consider 'design'. She bought a copy of *The Poultry Times* – which does actually exist – and trawled the web, visiting sites with annoying names such as One Fell Coop and Cock of the Walk. Pressure was applied, and not so gently. During bedtime stories she read 'The Emperor and the Nightingale' to Older Boy, who was then three. She emphasised the exquisite beauty of the gilded birdhouse the 'lucky' nightingale had. I responded with 'The Three Little Pigs', accompanied by admiring gasps at the house of straw and shouts of amazement at the house of sticks. She painted a better picture.

The Beloved has always had an inflated notion of my building prowess. She imagines concentrated industry, focused sawing, pencil behind ear, a slight, easy ripple of muscle as I fix joists and struts – she has a rich fantasy life and she cherishes this vivid, though largely unfounded, image. She has seen the results of my efforts in carpentry: the toy box I fashioned for the boys' room; the new doors I hung on the piggery; the table I made for a neighbour's nursery. What she tends to ignore is the actual process: hammered thumbs, twisted and drunken joints, anguished sobs, sliced fingertips and curses that would make a Marine drill sergeant quail. However, hope triumphed over experience yet again and I decided to give it a go. It didn't have to be too large, just big enough for half a dozen fowls at most. After all, we just wanted a few fresh eggs and for this we only needed a few layers. Just as there are different breeds of cattle depending on whether they are for beef or dairy, so there are hens bred for laying or eating – layers and broilers. We weren't planning to rear hens for the table at that stage because to keep up a consistent supply of broilers we would have needed a cock. Having a proud and handsome cock would have given me limitless opportunities for sad puns, but the price would have been too high. Younger Boy was ten months old at the time and he was subjecting us to his very own sleep-deprivation programme. The thought of voluntarily giving him an ally such as a rooster crowing intermittently but lustily beneath our bedroom window disconcerted us both. I might exhaustedly reach for the Dozol and the shotgun simultaneously and it would have been awkward to blearily confuse the recipients of those separate remedies. So it was settled – no cock for us (it's irresistible, really), just a moderate chicken run for a few chaste hens.

The great thing about building something to your own design is that there are no plans to misread and get wrong. There is no need to measure dimensions precisely and you are not subject to the tyranny of symmetry. You just make a bit, then you make a bit more to fit the first bit and so on. It is an organic process and a construction gradually evolves. I am still proud of the hen house that emerged and the Beloved was very pleased. It has two doors, one of which still opens and closes easily – a small one for the hens and a large one to allow us access for cleaning out the sawdust, chicken shit and straw, which make great additions to the compost. It has an almost square hole for interchangeable windows – Perspex in winter and a metal grille for ventilation in summer. It has four laying boxes at the back, just underneath the built-in hatch which enables us to remove the eggs with minimal disturbance. It also has a sturdy perch for the hens to roost on. The fact that this is positioned so that they crap directly into their laying boxes is a minor design fault. A major design fault is that the base is not really strong enough to hold the total weight and consequently the whole thing now sags drunkenly. But I am not going to be too hard on myself: foundation problems have often dogged the best architectural projects – look at the leaning tower of Pisa, for example.

The hen house itself was only the centrepiece of the planned complex – the chickens were going to need room to scratch around safely. So a predator-proof fence would also be needed. I sank tall poles around the perimeter and dug trenches between them. I filled these with concrete to set the chicken-wire fencing into and thus stop foxes and rats digging their way in. I even stretched more chicken wire over the top of the whole thing just above head height – there are buzzards in our valley and you can't be too careful; although seeing one swoop down to snatch a hen would have been pretty cool. However, after all my work, I would have felt rather foolish if my avian fortress had been invaded from above. As it turned out, the attack came from below.

We had a gala opening ceremony on the Beloved's birthday. She cut the ribbon I had festooned across the entrance and declared herself thoroughly satisfied. A few days later the first hens arrived. Older Boy named them Whitey, Brownie, Goldie and Biggie – he called it as he saw it – and before long they were laying and we stopped buying eggs. Absolutely fresh eggs really are delicious to eat and anything baked with them seems to be lighter and tastier. The hens were doing their job, and we did ours. They were well looked after. Apart from their regularly topped up grain dispenser, they had any left-over potatoes, rice or pasta and as much stale bread or toast as they could manage. They also scratched around for worms, beetles and bugs. We started to leave the door to the run open to let them roam around the acre. Once or twice they dug up tender seedlings and I had to chase them out of a vegetable bed with enraged shouts of 'Bloody hens!' But to be honest, I quite liked the constant contented clucking as they pecked around me whilst I weeded the peas and beans. They weren't too fat considering how much they were eating, but they were sleek, happy and hugely productive. Visiting city friends were especially pleased to leave with a jar of chutney and fresh eggs. All was well.

I was in the city rehearsing a Tennessee Williams play when I got an anguished call from the Beloved. She had been looking out our bedroom window, admiring the blossoming courgettes and the soaring sweet corn, when her eyes rested on the white-and-green hen house bright in the morning sun (she can be very lyrical) and she noticed something odd. The ground around it seemed to be writhing. She realised to her horror that it was a swarm of huge rats helping themselves to the hens' grain. The next morning she was at the nearest hardware store buying rat poison and the next Friday evening I was sitting under the silver birch tree with a gun. My searching gaze swept the immediate environs of the hen house. I was still, silent and intent on murder. I had borrowed a rifle from Male Farmer Friend and he had given me some cartridges specifically for shooting rats. They had a head of compressed sand rather than the usual metal bullet and were designed to be lethal to a rat at short range, without the possibility of a ricochet and inadvertent damage. I sat in the twilight, waiting. I was wearing a tweed cap, my leather jerkin and my special shooting gloves. These have a little hole near the end of the trigger finger and a little strip of Velcro, so that the tip can be fastened back to facilitate firing ease. They're cool! I looked up to our bedroom window and saw the Beloved looking down at me. Her face was contorted in a strange grimace. She had been quite shaken after her sight of the rat horde and she was still a little nauseous. I guessed she was anxiously anticipating the bloody rodent slaughter I was intending to wreak. I looked again and realised she was laughing – at me. I sat it out till dark. Not a rat appeared and it was getting chilly, so I went in for my tea.

From then on we scrupulously kept the hen-house door shut and the poison boxes full, but the rats kept appearing sporadically. I examined the perimeter fencing and found no evidence of a break-in. Then one day the penny dropped. The chicken run is close to the hedge that runs most of the way around our acre. On the other side of that particular stretch of hedge is a drop down a sheer-sided bank which borders the road. The rats must have been nesting in the bank directly under the hen house and digging straight up to feast themselves at our expense. The clever little feckers! I explored under the hen house and, sure enough, there was a tunnel going straight down. I got out the hosepipe and fed it into the hole, then turned it on full and waited by the bank with my gun – if any of them tried to escape I would shoot the bastards. The water ran for about an hour and nothing else really happened. I calmed down, put the gun away and turned off the tap. I must have drowned a few or at least encouraged them to move because there was an appreciable decrease in their depredations. There have only been a few isolated rat sightings over the last few years – I suspect the Beloved has poisoned many and I have shot one or two. On the other hand, quite a few chickens have come and gone.

The first to go was Goldie. We were all quite fond of her. She was the best looking. She had bright tawny plumage and feathered legs – she appeared to be wearing furry trousers. We first noticed that something was up when Younger Boy started to catch her regularly. Being a toddler at the time, it amused him deeply to chase the chickens and it amused us to watch him. The hens were unperturbed by his game. They just nipped out of his reach when he lurched towards them. Once in a while he would manage to grab one. He would whoop ecstatically and the hen would cluck tolerantly and submit to his affectionate hug until he let it go. Gradually it became clear that he was catching Goldie more frequently than any other chicken. He was doubtlessly speeding up, but she was definitely slowing down. Then she stopped emerging from the hen run with the others when they came out for a general peck and scratch. We examined her and found that her breathing was becoming a gurgling, laboured wheeze. After taking advice we realised she was incurably ill. We warned the boys that she might not last long and they said their goodbyes before going for a walk up the hill with their mother. When they had gone, I went to the hen house, gently picked her up and carried her out. She was rasping badly. I walked down the field and murmured soothingly to her. I think I even said something vague about my gratitude for all the eggs she had given us. Then I wrung her neck. When I held her upside down to do this, a generous cupful of fluid dripped out of her beak. She

was slowly drowning. It was the best thing. I threw her over the hedge and the next day she was gone.

Goldie was the first hen I throttled, but was certainly not the last. One summer we raised half a dozen broilers. We bought them as chicks from a poultry mart in a field in the neighbouring county. We had been assured that they would fatten up nicely and we brought them home in anticipation of plump roast-chicken dinners. The boys are annoyingly picky eaters and roast chicken is one of the few meals we can all sit down to with pleasure. We told the boys what the plan was for these chicks and they accepted it – they never got named and we reared them separately from the others. Their intended life span was to be comparatively short and we kept them in the piggery – which was empty of wood, it being the warmer months. We fed them well on the appropriate grain but they never got fat. They were healthy and vigorous, and they certainly grew, but they remained scrawny. We began to suspect that we had been diddled.

Female Farmer Friend called in for coffee one day and confirmed that we had been sold suboptimal layers and not juicy broilers, so the Beloved decided that I should 'do the job'. Older Boy was at school but his brother was still at home, so I made sure he was indoors and occupied before I began. He is a country boy and has a gruesome relish for blood and gore – he once cackled when he saw me chop the head off a shot woodcock – but I thought it best if he didn't actually see me kill the 'broilers'. I had dispatched two and laid them on the bank behind the house when the boy wandered out to say hello. I halted the slaughter. He toddled over to the dead hens lying in the sun, patted them gently and said, 'Wake up, henny, wake up, wake up.' I admit to a twinge of guilt. After a few more prods, he got bored and went back inside. I continued what I was doing.

It took some time to kill, pluck and draw them ready for the freezer and I wondered if it had been worth it. There wasn't a huge amount of meat, but when we finally ate them their flavour was rich and intense. It was the autumn of that year and we were having a bunch of friends for dinner, all of whom liked good eating. Most of them grew vegetables or had chickens or raised pigs: everyone squeezed around the table appreciated the ecology and especially the economy of home-produced food. We gave them onion soup made with our own onions. Then we ate the chickens in a delicious coq-au-vin with our own mashed potatoes and carrots. The pudding was a blackberry pavlova – the blackberries came from the hedge and our own eggs made the meringue. The only appreciable purchase for the dinner had been the wine for the chicken dish. Something I have noticed about growing your own food, particularly if you have raised and killed the animals that supply the meat, is that every meal is memorable.

Scaring Pheasants

Chickens don't live very long. Battery hens are culled after a year and a half, and a domestic bird is old at five. Our layers seemed to average about three years. As one or two went their way, we would get one or two more. The boys always named them. We have had White Tail and Black Tail; Donkey Kong and Princess Peach; and the most recent additions were called Rover, Spike and Butch. Possibly this last trio were named to give us a hint about what sort of animal the boys would really like us to get. Unfortunately, these designations failed to impart any canine qualities to the chickens concerned because they were totally unable to defend themselves last summer when the fox got them. The hen house has been empty and desolate since then. I was looking at it this morning. The November leaves lie in sodden layers on the roof, the weeds are colonising the run and the surrounding fence has caved in. I have to address this job soon.

Rover, Spike and Butch were snaffled before dusk. All that remained of them were large clumps of feathers in isolated corners of the field. We heard and saw nothing. Normally hens are safe enough roaming around in daylight, so we never bothered shutting them into the run until dusk. The fact that they were taken before dark suggested a very intrepid fox. Female Farmer Friend said there had been a marked increase in their numbers over the year. Consequently there has been increased competition for food and what we thought of as a daring fox was probably a desperate one – hunger has driven them to more and more reckless forays for food. Despite all the usual precautions a lot of the year's crop of lambs have been lost. A ewe with a single lamb is quite a formidable opponent – she could easily kick a fox to death and would eagerly do so to protect her offspring, which doesn't sound at all sheepish. All the ewes with multiple offspring are put to grass in fields lower down the valley where foxes seldom venture, however tempting it might be – it being much easier to separate a lamb from its mother if she has more than one to defend. But this year – foxes have ranged further down the hill than usual and taken more lambs than is acceptable, so Farmer Friends hired a marksman with infra-red sights on his rifle to sit out in their fields at night and protect the flock.

Despite this proliferation of foxes, the local hunt still can't seem to catch any. They are terribly enthusiastic, but they don't appear to be hugely skilful. They never seem to see a fox, let alone corner one. The Beloved went out with them last week and had a great time. Someone lent her a horse for the day and she borrowed a smart hunting coat for the occasion. She looked terrifically sexy. Afterwards she confirmed my suspicions – not a sniff of a fox all day, but that was beside the point. I am realising that the purpose of going fox hunting is not actually to hunt foxes: it is to have a mad gallop about the place, ideally with a lot of jumping and hopefully with a few scary moments where you nearly fall off but just manage to cling on and give yourself a bit of giddy excitement to gabble about in the pub afterwards.

I have changed my position on fox hunting. I used to be steadfastly opposed to it, now I don't give a fart. The England I lived in as a young adult was a bitterly divided place. There was a very decided political demarcation between two hostile parties. Once you knew which side of the ideological chasm you inhabited, and it was pretty clear to me where I belonged, you inherited a whole raft of opinions on every question. I was against hunting because it seemed to be the preserve of a rich rural elite; it reinforced class barriers and it entrenched the town and country divide. My concerns about animal welfare were negligible – even at the time the fury directed against fox hunters on these grounds seemed disproportionate. It was like being against the arms trade and pointing the finger at a manufacturer of penknives. Millions of battery hens led abysmal lives and hunts killed perhaps hundreds of foxes … or none, if you live around here.

Our local hunt does not seem in any way exclusive – most families seem to have somebody involved, regardless of social position – and class distinctions are vague here anyway. There is an apt phrase which is commonly used to puncture social pretensions: 'Everyone is only a few generations from the bog.' The hunt certainly has one or two posh-sounding members, but the current master has a broad Wicklow accent and anyone can join if they have a horse, which covers pretty much everybody roundabout, apart from us. At larger meets I have seen glossy hunters, carthorses and hairy nags set off after the hounds. When I am a decent enough rider and can borrow a mount I might give it a try myself. I already have the hip flask – now I just need the rest of the gear. I have no concerns about ever having to witness the hounds make a kill. I see a lot of dead foxes on relatively quiet local roads – this suggests our native ones are not the most cunning exemplars of their species, and the hunt still can't catch them.

The fox population boom seems to be widespread. Last weekend I was chatting to the gamekeeper of a large landed estate in the north of the county. He was deploring the increase in fox numbers because they threatened his pheasants. He has been observing them in daylight, which is rare enough, but he has also noticed strange, untypical behaviour in some. He thinks these are probably city foxes that have been pushed out of the suburbs and into the country. Urban foxes seem to be increasing too, and his suspicion is that the young ones are striking out to find their own territory and food supplies – vulpine economic migrants, I suppose. The gamekeeper is doing his utmost to deter them. He prefers not to trap them, as he considers it inhumane, but he will go lamping for them after dark. It is part of his job and he is well aware of the ironies of his occupation.

A year or two ago, we were standing on a heather-covered bluff looking across the estate. The morning was crisp and bright and the blue sea was twinkling over the shoulder of the purple hill opposite. Above our heads a brace of cock pheasants were flying high and straight in the clear air. They had just been flushed out of the woods behind us and over the line of guns in the valley below. He said, 'I spend the year raising these birds from chicks. I protect them from predators, feed them and nurture them to a peak of fitness, just so that people can shoot at them.' We heard five or six shots from the guns. 'And miss,' I added as the birds disappeared over the hill. He shrugged and grinned. As I followed him down the hill, I envied his boots. They were knee high and made of sturdy but supple leather, with great grips on the soles – like really superior wellies. The rest of his outfit was a mixture of more leather, heavy corduroy and waxed cotton in muted browns and greens – functional but cool. And he had a terrific hat. He is also rugged, dark and handsome – an idealised Mellors to the life. I must ensure the Beloved never meets him.

The birds he cares for are to supply a 'driven' shoot. When I try my luck walking with my gun in the fields around us, it is what is called 'rough' shooting; a driven shoot is more organised.

The ‘guns’ are arranged in a line of marked positions, or ‘pegs’, at a strategically chosen spot – say, for example, in a field bordering a wood. At a prearranged signal, in this case a single blast of a horn, a ‘drive’ begins. A line of beaters will start to slowly approach the guns through the wood making a steady, sustained noise – clapping hands, banging sticks or beating bushes. It is particularly useful to have gun dogs along to flush out any birds that the beaters might miss. The combination of dogs and noise theoretically drives the pheasants towards the line and should ideally ‘put them up’ over the guns, although this seldom happens. Two blasts of the horn signal the end of the drive and firing stops. It is an approximate science: too much noise and the birds freeze or ‘sit tight’ and don’t move from their cover; too little and they just scurry back, unseen, through any gaps between the beaters.

It can be a little disorganised at times, but that is part of this particular shoot’s attraction. It is genial and relaxed and I try to go along once or twice each year during the shooting season. I have heard of other competitive shoots where the gamekeeper is a martinet and screams at beaters to ‘hold the line’ and where inaccurate shooting is sneered at. This sounds deeply unattractive. There is a commercial shoot not too far from where we live on what used to be an aristocratic estate. People pay huge amounts of money for a day’s slaughter. The birds are very professionally funnelled directly over the guns’ heads and are butchered all day. I suppose it is a fine distinction, but at the one I attend the shooting seems much more ‘sporting’ and the measure of fun had is not solely determined by the size of the ‘bag’ at the end of the day. The estate has been in the hands of the same titled family since the early seventeenth century, although it is now a fraction of its original size. There are the usual public attractions employed to keep such places going: tours of the house and gardens; a weekly farmer's market; arts festivals and so on. It has also been used as a location for film and television projects, although I have yet to shoot anything there, in that sense of the word. However the pheasant shoot is not for profit and makes no contribution to the running of the estate. It is financed by a syndicate whose annual subscriptions pay Rugged Gamekeeper's wages and cover the upkeep of the pheasant population. For this they get ten days' shooting a year. It is not exactly cheap, but it is substantially less expensive than the fees extracted elsewhere. The earl allows the use of the land in return for a bit of shooting for him and his family, and everyone gets a few pheasants. A friend from a neighbouring village who has a few bob is a member of this syndicate, and three or four years ago he asked me if I wanted to come with him for the day. So I tagged along.

When I rose early last Saturday morning to tag along again, the eastern horizon was red. Sliabh Buí was maroon, and the lurid pink sky was streaked with carmine clouds. It was a gaudy 'rosy-fingered dawn'. We have been enjoying what used to be called a 'Saint Martin's summer'. This is the old name for a spell of mild weather around the saint's feast day in the middle of November. The general consensus in the village is that it has been great weather for this time of year. In fact I heard someone remark that 'It isn't this time of year at all!' The fact that there is an old name for warm weather during this month suggests that it has not been uncommon in the past, yet it still surprises us.

My shooting gear is not a patch on Rugged Gamekeeper's, but it is warm, weather proof and appropriate shades of green. I threw it in the car and drove through the old oak forest to Got a Few Bob

Friend's house. By the time we got to the estate the sun was fully up. We turned off the main road and down the long beech-lined avenue. Passing through the wrought iron gate, surmounted by the family crest, we pulled up by the lawn in front of the house. Most of the current building is in the Elizabethan revival style of the early nineteenth century and its castellated facade makes an apt setting for the gathering. There are about twenty members of the syndicate, mostly men, but there are a few women shooting too, and spouses, children and friends come along for the day. And there are always plenty of gun dogs – setters, spaniels and Labradors are typical but occasionally more exotic breeds like Weimaraners appear, along with a few mongrels and mutts. Everybody dresses in muted country colours and in fabrics you would expect – tweeds, waxed jackets and thick cords – but there is also a lot of the paraphernalia peculiar to shooting: different styles of cartridge belts or pouches; leather or canvas gun slips; a wide variety of caps or hats; and three-quarter or full-length waterproof chaps – these are great for keeping your legs protected and dry when you are bashing through sharp gorse or damp ferns. I must get some.

A day's shooting always follows the same basic pattern, and last week's was typical. People mingled in loose groups around the lawn, chatting affably and ensuring their dogs didn't foul the well-kept grass. The earl wandered amiably from group to group, welcoming newcomers and greeting regulars. He is a genial Irish gentleman with a broad smile and impeccable old-school manners – he always doffs his hat to a woman. Got a Few Bob Friend's dad is one of the 'captains' of the shoot (the earl's brother is the other) and he called us all together to introduce guests and any new 'guns' who may have joined the syndicate. Then

Rugged Gamekeeper took over. He went through the essential safety precautions, as he always does: never carry your gun loaded; always unload your gun before passing it to someone and always hand it stock first; and never fire unless you can see 'blue sky beneath both barrels'. He added that we should avoid 'ground game', by which he meant rabbits, foxes and hares, and he particularly emphasised that the countess's ducks should absolutely not be shot. Apparently it enrages her.

The guns were then divided into two teams – team A to shoot the first drive and team B to help beat it, and vice versa for the next one and so on throughout the day. Then we set off. The guns had a fair way to go so they were piled into a trailer towed by a Land Rover and taken to their pegs. The rest of us walked by the side of the house, past the glass-domed orangery and along the side of gardens planted in the seventeenth century. The earl told me a little about the size and history of the estate as we walked. I had noticed varying degrees of stylishness in the invariably green shooting outfits on display – some of it very smart and new – but his gear was extremely worn and verging on the shabby. Well, it was his house and land – he didn't need to impress anyone. We climbed a crag with a beautiful view back over the house and gardens, waited for the single blast of the horn and then headed into the woods to begin beating the first drive. I am a much better judge of the relative impenetrability of undergrowth than I was. The first time I helped beat this wood, I plunged keenly into the densest spots. I got completely tangled in impassable briars and whimpered with each thorny cut as I struggled to get out. When I finally escaped, red faced and sweating, with my tweed cap annoyingly askew, the earl's brother exited a nearby thicket in similar disarray and we nodded manfully at each other.

I compared our circumstances to those Japanese soldiers who finally emerged from the jungle years after the war was over. He barked a laugh and dived eagerly into the next bush and I felt obliged to follow.

These days I know enough to leave the really hard stuff for the dogs and so I chose a path with less resistance. As we walked in a vague line, the pheasants started fluttering up from their cover, the beat of their wings making a very distinctive whirring noise. They hooted as they flew off over the treetops pursued by shouts of 'Over!' or 'Forward!' or 'Back!' from the beaters – this is intended to give the guns some warning of the direction a bird may come from but I am not sure how helpful it is. I suspect the guns just heard a confused babble of excited bellowing, whilst the beaters mostly heard a confused fusillade of excited shooting. The birds mostly scattered in all directions but a few were bagged before we all reconvened for the second drive.

This was some distance away so everyone – beaters, guns and dogs – squeezed into a couple of trailers and we bumped along muddy tracks till we pulled up on the shoulder of a high hill. The two teams swapped over and the new guns went to their pegs in the valley while the rest of us began to beat down the hill towards them. We flailed through tall, dead ferns, but they were easy enough to pass. I didn't 'put up' any pheasants on this drive; however, I did flush a deer. It was a Sika stag. It hurtled out of some gorse just ahead of me and flashed past a yard to my right then disappeared into the woods behind: a brief but beautiful glimpse. When this drive was over, we all went back to the front lawn for a picnic lunch. Everyone brings their own food and drink but there is always a bit of bartering. I swapped some of my homemade mushroom soup for some meat loaf, and a bottle of cold beer for a cup of hot coffee. I lit a cheap but delicious cigar and nattered to an inner-city Dubliner who grew up near my granny and then chatted to an Austrian count about Viennese writers. This quite mixed group of people is united by common conviviality and, of course, a shared interest in shooting.

There is a wide range in the level of skill displayed, but being a good shot doesn't always mean a big bag. Luck also plays a part – the pheasants either come your way or they don't and you don't take a shot at a bird that passes someone else's peg, regardless of how many, or how few, have come your way. Someone shooting well is a source of general pride and someone shooting poorly elicits commiserations and, it has to be said, occasional laughter. There is still amusement to be had from a story of someone unnamed taking a few unsuccessful shots at a pheasant roosting on a telegraph pole and completely destroying the white ceramic knobs connecting the power lines. The Electricity Supply Board was furious, and the earl was urbanely annoyed.

After lunch the two remaining drives had a more leisurely air. For the first one, the pegs were arranged in a long line at the edge of a beech wood and up a gradually sloping field which had produced barley over the summer. I was at the bottom, looking up at the guns waiting in the stubble. Their shooting greens blended into the landscape. There was the occasional puff of smoke with a delayed report of a shot if a pheasant emerged from the treetops. The afternoon light was turning the top of the distant brown hill golden. It was a sight that could have been seen here for more than a hundred years – then a double-decker Dublin bus chugged along the road at the top. The last drive of the day was in sight of the house and Got a Few Bob Friend let me shoot his gun for it. I missed two coming up out of the woods directly in front of me – they were high, but still within range, and a better shot would have got them easily. Then he alerted me to a bird popping up out of a large bush well over to my left. I was the last gun at that end of the line, so I whipped around and got it over my left shoulder as it was flying away. I seem to be a much better shot when I don't have too much time to think.

The double horn blast sounded for the end of the shoot and Got a Few Bob Friend and I walked back towards the house along the gravelled border of a long lake. Unharmed ducks were quacking on the water and an ornamental fountain was playing in the fading light as we made our way to the gun room for tea. It was crowded and noisy when we got there. It is in one of the outbuildings attached to the old stables behind the main house. It has a flagstone floor, a high raftered ceiling and a big open fire. Around the white limed walls hang an old map of the county marked with where particular game is common and a long list of shooting dos and don'ts written up a century ago. Interspersed with these are a few sets of deer antlers. We drank a welcome mug of hot tea and wolfed sandwiches and cake – after all, lunch had been ages ago. The bag for the day was just over a hundred pheasants, so, with twenty guns and a little more than that in friends, beaters and spouses, everyone took home a brace.

I wonder how long the shoot will survive. Responsibility for the estate is shifting to the earl's eldest son, who has been developing new ideas, the farmers' market amongst them. This is spreading and growing and as it does , the gun room is being used as a break room for the stall holders. As Christmas approaches, the old stables have been cleaned up and turned into Santa's centre of operations, with workplaces for elves, reindeer stalls and plenty of parking for sleighs. It is lovely, but not exactly commensurate with what is after all, a blood sport. So things will change and there are mixed feelings in the gun room. Got a Few Bob Friend and I said our goodbyes, grabbed a brace of pheasants each and headed home. I hung mine for four days before plucking and drawing them. They are now in the freezer along with a rabbit or two and quite a few pigeons. I am thinking of attempting a raised game pie for Christmas.

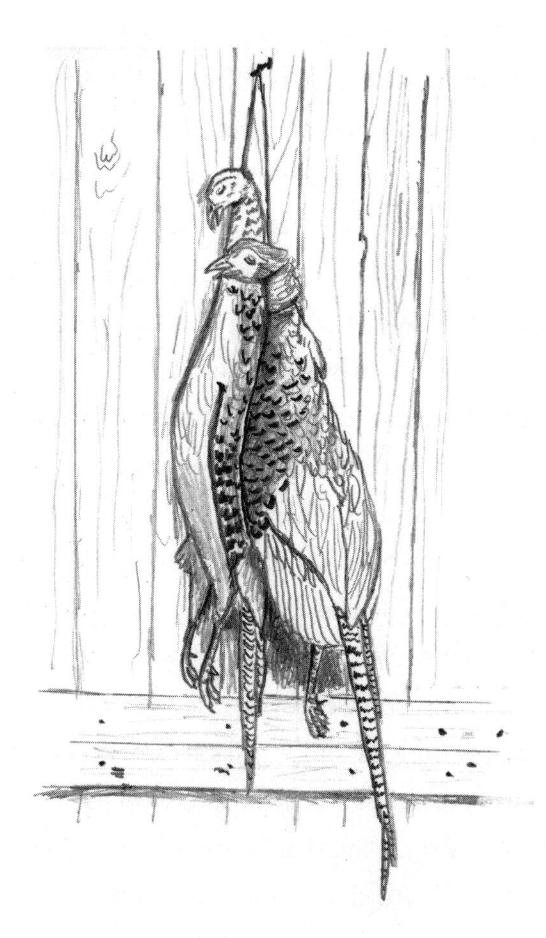

PHEASANT CURRY

Ingredients

a brace of pheasants
2 medium onions
3 hot red chillies (or what you can handle)
2 hot green chillies (as above)
1 golf ball-sized piece of root ginger
4 garlic cloves
2 tbsp ghee, coconut oil or sunflower oil
pinch of turmeric
pinch of ground nutmeg
1 tsp celery salt
2 tsp ground cumin
seeds from 20 cardamom pods
2 cinnamon sticks
2 x 400ml tins coconut milk
a large bag of baby spinach
salt and pepper

Method

Remove the legs from the pheasants and separate the thighs from the drumsticks. Carefully take the breasts off the bone and cut each in two. Put to one side.

Roughly chop the onions and chillies and put them in a food processor. Peel the ginger and garlic and add. Blitz to a paste.

Heat the oil in a large saucepan or casserole dish. The pan needs to be big enough to freely stir the contents. When it is very hot, add the onion mix and pheasant pieces. Stir until it starts to take on colour. Mix in the turmeric, nutmeg, celery salt, cumin, cardamom seeds and cinnamon sticks. Season with plenty of salt and freshly ground pepper and cook for another minute or so. Turn down the heat and pour in the coconut milk, not letting the heat rise above a faint simmer. Cook, lid on, for one and a half hours or until the sauce has reduced to a rich consistency. You might need to take the lid off to reduce the sauce to your liking. Five minutes before the end, add the spinach. Serve with basmati rice, a handful of chopped coriander and some of your favourite delicious chutney.

NOLLAIG

Biting Frost

Hanging Tinsel

Sliabh Buí is invisible beneath a grey shroud of rain, the woodpile in the piggery is dwindling and money is tighter than a python's coils. And the day didn't start well either. Normally there is a general lack of urgency on Sunday mornings, as we attend neither church nor chapel – but not today. It is the first day of December and the boys were excitably awake for two reasons. Primarily they were impatient to sample the first chocolates from their Advent calendars, but also they were eager for us all to get to the church in good time for their Christmas concert. I wasn't quite ready to spring out of bed when they bundled into our room a little earlier than expected. I had sat up quite late the night before on the bench at the front of the house, gazing at the dark fields. It was a clear night and the stars were bright. Sliabh Buí was a solid black mass. I drank some wine and smoked a cigar as I waited for shooting stars. I do this often. It is solitary, peaceful and soothing. It helps to erase anxiety and lets me dream – usually of finer wine and better cigars. I enjoyed my own hospitality for a while, and by the time I came to bed Orion had moved over to my right and the moon was up. I heard the rain begin to patter on our roof well before dawn and didn't sleep deeply after that. So when the boys piled into our bed – all sharp elbows and butting knees – it wasn't as welcome as it used to be.

They call this 'having a snuggle', which is what it was … when they were much, much smaller. These days it's more like tag-team wrestling as they squeeze in between us, grunting and jockeying for position. We need a much bigger bed. I have become quite skilled at balancing on a four-inch-wide strip at the edge of the mattress and snatching a few more minutes' rest, but not today.

When the Beloved started elbowing me in the kidneys with a gentle whisper of 'Tea,' I surrendered and got up. I put the kettle on and set the stove. There was enough kindling and logs to get it going, but the ash tray under the grate was full and needed emptying so I had to brave the weather. There was still a slight rain falling but it was mild enough. I slipped on my Crocs and nipped out to the ash bin behind the piggery – it was full to overflowing. I trotted through the sodden grass to throw the ash in the hedge, taking great care to keep the tray downwind so I wouldn't get plastered with its contents. As I was emptying it, a sudden gust whipped most of it in my face and down my pyjama legs. I swore filthily.

When I reappeared in the bedroom, heroically bearing juice for the boys and tea for the Beloved but also smeared with ash, I was denied access to the bed. I came downstairs and laid the table for breakfast, loudly, with much disgruntled clattering of cutlery and plates. This wasn't too good for my head, so I sat down, drank a pint of tea and contemplated the Christmas concert with a jaundiced mind. It was far too early to be feeling festive, although the boys had recently been in frequent, solemn conference over Santa Claus. This isn't necessarily seasonal – he is a fascinating subject to them at any time of the year – but we had otherwise managed to avoid the general gearing up for Christmas.

Some time ago the Beloved and I decided to preserve our mental health by switching off the radio which used to play all day in the kitchen. Listening to the daily barrage of corruption and incompetence in public life, and the insouciant greed with which the financial junta continues its tyranny, was depressing her and infuriating me. When we watch the television it tends to be digitally delivered backlogs of much-lauded drama that we missed first time around, so we have mostly avoided the advertising onslaught which renders Christmas inevitable and inexorable. But now I had to face the reality of its imminence and its massive expense. My calculations were not concerning how much we would have to spend: they were about how much we would need to borrow. Scrooge's 'Bah! Humbug!' seemed pitifully inadequate. I looked out the window and noticed it had stopped raining, so I went outside to survey the kitchen garden and think about planting garlic.

Now is a good time to do this. If the shoots are about three inches high when the late-January frost strikes, it helps to split the garlic heads into cloves, so planting about a month before the winter solstice is ideal. For some years now, I have meant to do this on the first day of Advent, thinking it might become a nice and easy annual ritual. I looked at the unweeded garden grown to seed, possessed by things rank and gross in nature, and grew discouraged. For the time being, the planting of the Advent garlic will remain a custom more honoured in the breach than the observance. I came back inside and made pancakes.

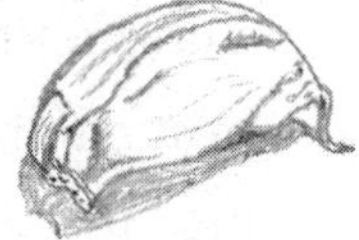

The boys were in high spirits when we dropped them at the school just across the road from the church. They were eager to give their concert, and having started the day with a chocolate figure from their calendars, they were particularly ebullient. I was feeling less so as we took our seats near the front, the better to get a good view. The lights suddenly went out and the church was in darkness, then they came back on. There was much shuffling from the back of the church, then some fierce adult whispering, then a pause, then the lights went out again and the children came in. They processed slowly up the aisle two by two: each carrying a small electric candle and all with serious, intent expressions on their small faces. They sang a sweet melody in Latin called 'Deus Meus' as they made their way to their pews. The vicar handed out oranges wrapped in tinfoil with cocktail sticks sticking out the side and a candle in the top. Apparently this is a 'Christingle' – a Christmas totem which originated in the Moravian church in the eighteenth century: the orange is the world, the cocktail sticks symbolise the seasons and the candle is the light of the Christ child. I suppose we must be represented by the tinfoil. Then the children took their positions in front of the altar and sang at us. One little girl in a red dress, with severely brushed long, dark hair, had a stricken look on her face. She was overwhelmed with the awe of the occasion and had to be helped back to her seat when the singing was done. The little girl next to her in a white dress, with severely brushed fair hair, yawned throughout. During the readings, Older Boy delivered his piece with clear diction and nice resonance. We were very proud. During the last song, I was watching Younger Boy closely. I had the words in front of me on the printed order of service and he seemed, by the movement of his lips, to be singing something else entirely. I hoped it wasn't his rude version of 'Jingle Bells', which he bellows at the least provocation. I scrutinised him carefully, he glared emphatically back and I realised he was mouthing, 'Don't look at me!' The final number was the well-known Largo movement from Dvořák's 'New World Symphony' played on massed penny whistles. I left the church feeling much more benign.

This slightly cheerful, festive mood was only partially eroded by a day's Christmas shopping in the city a week or two later. When I lived in town I did my Christmas shopping in sporadic bursts of enthusiasm over a period of a few days. But not any more – I usually have just one day to do it, like country dwellers of the past. This was known as going to *margadh mór* or 'the big market'. People from rural areas would come to the towns with eggs, poultry, vegetables and dairy produce. They would return with spices, sugar, tea, children's gifts and drink. These days all of these items are, of course, readily available hereabouts, especially the last. But I still like to visit the city to look at the lights and purchase esoteric things like books and smelly cheeses. I drove up to town the evening before to give myself a full, clear day to attack the shops. I was cautiously confident that I could make my limited funds stretch as far as I needed them to. Imagination and improvisation would ensure success. I just needed to be clear headed and open to inspiration. This plan was sabotaged by going out to dinner that night. An old friend and colleague from my London days is on location here for a few months, and we had been meaning to catch up. She is earning and therefore, according to a time-honoured tradition of the profession, she insisted it was to be her treat. I chose a restaurant conducive to senseless roistering and we lived up to our surroundings – dinner was delicious, the wine gushed and gurgled and so did we.

Unfortunately, I am out of practice with such nights and the next morning I thought I had been poisoned. I felt, as my grandfather used to say, 'like a sackful of arseholes'. Negotiating the crowded streets and shouting shops took grit, determination and regular stops for coffee. I had a list – the Beloved has taught me the value of these over the years – and I clung to it like someone in a hurricane clings to an umbrella. It was only slightly more effective, and I drove home after dark with less shopping than I had hoped for. As I gripped the wheel grimly, I had a brief aspiration that Christmas had come and gone in my absence. This proved to be a state of mind impossible to sustain.

The last fifty minutes of the journey home is a winding drive along unlit country roads through a handful of villages and one small town. At this time of the year you pass a few sporadic houses lit up like the mothership in *Close Encounters of the Third Kind* and, whatever my mood, they lift up my soul. On one particular straight stretch of the route, there are two adjoining but otherwise isolated bungalows. They must be in competition with each other for the most dazzling illuminations, and each December the stakes get higher. This year the spectacle is epic: both gardens are packed with sparkling snowmen, reindeer and angels; the twinkling trees flicker like multicoloured beacons; the glittering walls and roofs are crowded with abseiling or skiing Santas; and miles of fairy lights are draped wherever there is room to hang them. The whole effect is garish, trashy and magnificent. I laugh out loud whenever I pass, and one of these days I intend to pull over and leave a gift at each door with a note expressing my gratitude for years of pleasure. This time as I drove past, I slowed right down to admire the new additions. Despite a longish tailback behind me, nobody beeped in complaint.

We are indirectly involved in a Christmas lights competition of our own. Our village is one corner of a triangle of local settlements. One of the other two is slightly bigger than ours and one slightly smaller. The Beloved and I have a long-running disagreement about whether the bigger place is a large village or a small town. She claims it is a town because you can get a spray-on tan there, as if this clinches the matter. I say it is a village because as you drive into it from our direction you pass a sign which reads 'Please keep our village tidy!' I feel this makes my position irrefutable. The three villages ceremonially turn on their respective festive streetlights over a mutually agreed weekend in mid-December. This is done on successive evenings every year so that everyone can go to each other's. An impartial observer would find it difficult to judge the best, as there are three quite distinct styles on show. The small village has a huge solitary chestnut tree in the middle of the green. They drape this liberally with red, green, blue and yellow lights and there is a single white star at the top. Nothing else is illuminated and this accentuates the impact. The effect is one of simple, consummate beauty. The village is on the crest of a hill, so when you approach it at night and the tree is suddenly revealed, it seems to float in the darkness above you. It is glorious and gladdens the heart. The big village is a little more comprehensively adorned, but just as restrained. They erect a traditional pine tree in the square and decorate it with simple white lights. The surrounding houses have small spruce trees protruding at a slight upward angle from the walls. These are colourfully yet tastefully lit. There is an air of classic elegance. It is muted but very, very pretty.

In our village, the look we go for is inspired by the Las Vegas strip. The main street is festooned with strings of mismatched neon pinks and blues. And the lines of lights emanate quite some way in all directions outwards. We value quantity over quality and we achieve a mood of gaudy razzamatazz. I love it.

We turned on our lights last weekend to mixed emotions from the boys. The ceremony always begins with the arrival of Santa Claus. He turns on the lights and then everyone repairs to the larger of the two village pubs so that the children can meet him, score an early present and have a party with lots of junk food and dancing. It's hellish, really, but the kids usually enjoy it. There was much excited anticipation earlier in the day: Younger Boy had declared the three best people in the world to be me, the Beloved and Santa – I suspect this is in ascending order of importance; and he and his brother spent the hour or two after school speculating on what present they might get and how many packets of crisps they could eat before being sick. Expectations were high.

The previous evening we had all been to the small village for the turning on of their tree. Santa had zoomed up the hill on a motorised sleigh, which I thought a bit vulgar and showy and completely out of keeping with the temperate simplicity of their lights. The boys were also a little dubious – they seem to be quite traditionalist in their views. Our Santa approached slowly in a more conventional horse and cart. I looked at Younger Boy and nodded appreciatively but he was scowling suspiciously. I looked at Santa again and had to acknowledge that his costume was rubbish. When he shouted a 'Merry Christmas!' to everyone in a broad Wicklow accent, Younger Boy snorted angrily. This Santa clearly didn't come from Lapland. Younger Boy was rigid with fury. He declared Santa to be a fake and demanded to know why we had brought him here. I tried to explain that of course this wasn't the real Santa and perhaps we should go to the party anyway. He remembered the possibility of a present and begrudgingly acquiesced. Unfortunately he got a truck. He thought this was for babies and howled like one until he remembered the possibility of junk food. The tears evaporated instantly and he headed for the buffet. He managed three bags of crisps, four chocolate bars and a pint of lemonade – so he ended the evening in a happy mood, even if it was more of a sugar rush than a natural high. His earlier disappointment was, of course, because nothing can live up to his hopeful imagination.

Three years ago we were watching the switching-on of the lights in the smaller village. The tree was ablaze above us, Santa's sleigh was twinkling beneath it and fireworks were bursting in the sky. I was holding the boy in my arms so he could see properly. He was a bit bewildered but his face was alight with amazed joy. He was three and a half at the time and I could see he was struggling to express something important, something nearly too vast and wonderful to comprehend. Slowly and almost stuttering he said, 'Christmas is everywhere ... Santa is everything.' That was half his short lifetime ago but the memory must linger still, and nothing subsequently is going to beat it for overwhelming happiness.

Older Boy was quite philosophical about the second-rate Santa – he is eight, after all, and worldly enough to tell a bogus Santa from the real one. He was also sufficiently pleased with his present and a little more restrained with the crisps and lemonade than his brother – then he hit the dance floor. This time last year I watched him hover shyly at the edge of the floor till he eventually plucked up enough courage to plunge in, dance vigorously for fifteen seconds near a girl, then retire sheepishly until he regained the confidence to try again. I observed him surreptitiously and my eyes moistened: partly for his own sake and his embarrassed, eager awkwardness; and partly for my own suddenly vivid boyhood being re-enacted before my eyes. This year he was more considered in his approach. He spotted a group of girls he is familiar with and he watched them carefully as they rehearsed some steps. Then he joined a girl he knew well and concentrated on his footwork. This time he lasted much longer before self-consciousness reasserted itself, and when everyone 'got on down' and fake rowed along the floor to 'Rock the Boat', he flung himself in enthusiastically. He ended the evening red faced, sweaty and delighted with himself.

We waited one more week before we finally legitimised the season and put up our Christmas tree. I miss being able to cut one of our neighbour's trees up the hill, and for years now I have been harbouring a plan to plant some saplings each January, so that in years to come we can cut our own. My current plan is to put the tree-planting plan into actual operation some time soon. This year we bought one from the village shop and Christmas sort of arrived with it, as it usually does. One December when Older Boy was still a toddler, he pointed at the just decorated tree and declared, 'Christmas … over there!'

A mood is often evoked by sensory associations and particularly so at this time of the year. The aroma of pine needles is strongest when the tree first comes indoors. I had just made mincemeat and the house was full of the warm smells of spices, brandy and suet. I dug out the Christmas music CDs and played something appropriate as the Beloved and the boys unpacked the decorations and lights. The concept of the Christmas CD is utterly naff, I know, but the boys like the tunes – the younger one's favourite is 'I Want a Hippopotamus for Christmas' and his brother favours 'All I Want for Christmas Is My Two Front Teeth'. My personal favourite is 'I'm a Little Christmas Cracker'. The Beloved rolls her eyes and pretends to be above it. I claim I am being ironically post-modern but I actually enjoy the kitsch humour. As Noël Coward said, 'There is nothing as potent as cheap music.' I will try to get away with such stuff as long as it is tolerated.

We positioned the tree in the usual corner. I am a little bit more relaxed than I used to be about its precise verticality, but the boys were scrupulous about placing it the correct distance from the walls. It emerged that this was to allow enough room to for them to hide behind it and eat stolen chocolate angels unobserved. It was Older Boy's idea, and Younger Boy was so dazzled by its brilliance he felt he had to tell us. I made the Christmas cake in the middle of November and have been feeding it with brandy since then. I got it down as the tree was being dressed and opened a bottle of something sticky to have with it. There was a time when I sourced obscure Liqueur Muscat from the Rutherglen Valley to drink at Christmas; these days Cream Sherry from Adl (or is it Lidi?) is perfectly adequate. The boys sat at the foot of the decorated tree staring up at it, and we watched them as we ate our cake and sipped our wine. They had lemonade and a candy cane each – they don't like Christmas cake, which is just as well as they are far too young to drink.

I am now quite ready for Christmas. The final festive factor – the icing on the cake, as it were – has been the icing on the fields. We had our first proper frost this morning. The winter barley across the road glittered in the yellow light of the early sun and Sliabh Buí was just visible above a blanket of mist. I left footprints in the frozen grass when I went to get wood from the piggery and my breath was visible in the sharp air. It had been a clear night and it looked like it was going to be a clear day so I thought I should replenish our diminished store of fuel. Male Farmer Friend had mentioned a few boughs that were down, so after the boys had set off down the hill to school, I got out Lesley the Lumberjack's outfit and headed off. There were no cattle about – they are now in the barns for the winter months – and the only sound I was aware of was my boots crunching wafers of ice on the muddy track. I climbed the gate into the pound field, where there were a couple of large ash boughs waiting. Twenty minutes with the chainsaw reduced them to manageable lengths and then I headed for the spring field where I found a sizable branch of oak. I dealt with that similarly and dropped in to the farmhouse to borrow the pickup truck so I could collect the wood.

Male Farmer Friend was having coffee. He asked me to join him and showed me a snapshot he had taken first thing that morning. However early I get up, he is always up earlier. It was a great picture. He had taken it from the top field behind our house, looking across the valley to Sliabh Buí. It was a familiar view but from a much higher vantage point. The mountain glinted above the fog, just as I had seen it do earlier, but in the foreground, and dominating the photo, was a handful of fine-looking sheep wreathed in the steam of their own breath. He was about to post it on the web and he showed me other photos of sheep from the same Twitter stream, all posted in the previous twelve hours. There was one from Saskatchewan in Canada: it was against a background of deep snowdrifts, and the thick layer of flakes on the sheep's fleeces remained un-melted – which shows how effective the insulation is. There was another from New Zealand, where it is currently high summer – this flock was a bright white patch on a rich green field under a clear blue sky. This stream is followed by sheep farmers around the world. Every photograph tweeted is beautiful. They all seem to speak of a sense of place, belonging and ownership. And there is a mutual respect for and pride in a shared occupation. I love this juxtaposition between traditional lifestyles and new technology. Over the hill from us lives an old farmer who sorely misses his son who had to move to New South Wales for work. Once a week the son straps his smartphone to the dashboard of his tractor whilst he is working a vast Australian field and Skypes his dad in Wicklow.

Emigration is once again a melancholy fact of Irish life. There was a brief respite during the Celtic Tiger years. In fact, not long after I myself returned, much was made of the fact that for the first time in decades – indeed it may have been for the first time since the Famine – more people were returning to the country than were leaving. Now the young are departing again. This is undoubtedly a sad thing for the individuals concerned, and for their families left behind. It is also a big worry for the local Gaelic football team, who have recently lost some of their best players.

The boys have been playing a bit of football on the village green for some summers now, and one or other of us brings them down on a Friday evening during the season. I have never been sporty and I haven't played Gaelic football or hurling – the GAA sports – since I left Ireland as a boy. I don't have a firm grasp of the rules so I would usually bring a book while the boys ran around the field in vague pursuit of the ball. Like all other parents, I was on a texting list to keep us informed of the upcoming Go Games. These are held every second Saturday morning. In successive villages in our general area, four or five nearby teams gather to play a bunch of matches of ten minutes each way. There are usually a few teams from each village, boys and girls together but separated by age into under sixes, under eights and under tens. It is all very easy-going and the emphasis is on fun – no score is officially kept, although the parents usually have a clear idea of the final result. The children enjoy it and they get to run a lot – often in the right direction.

About this time last year, I received the general text alerting parents to the AGM of the juvenile branch of the village club, and the Beloved felt I should go. I had no desire to do this – I pointed out that I knew little of sport generally and less than nothing of the Gaelic Athletic Association. She said, 'Your sons play! Go! Join in!' The meeting was to be in one of the village pubs and there is no harm in a convivial pint, so I went. It wasn't a particularly late night, but she was asleep when I got home. The next morning she chortled when she asked me how the meeting had been and thought it the height of wit to enquire sarcastically if I was now 'on the committee'. I replied, 'I'm the new vice-chairman, actually,' and she laughed for twenty minutes. I had been nominated, seconded and elected in under a minute. I did demur for a moment. I pointed out that I knew nothing about the game and that I was unsure of committing to any responsibility. I was assured that it was largely a 'ceremonial' position and that I 'wouldn't have to do much'. They knew how to sell it. I wondered if I might get a sash or something and thought of the Beloved's amusement – and the title. I accepted my election.

Sadly there was no sash and I ended up doing more than a little. I went to many meetings and spoke at some. On one occasion I even raised my voice emphatically, but I was drowned out by the general roaring and shouting which followed the unveiling of an extremely controversial team kit. I also helped out at the Family Fun Day during the summer, the memory of which is still tender. My especial responsibility had been to chalk a number on each plastic duck as it was purchased prior to being chucked in the river for the Grand Duck Race. Before the event, various committee members and numerous volunteers gave me lots of pieces of paper with the number of ducks they had sold, along with the names of the buyers. There were a lot of names, numbers and ducks. There were also lots of passers-by to chat too but I concentrated and got on top of it. Eventually, every duck had a number. But not every number had a duck. And some numbers had a duck but no purchaser. The fact that the total on the board, and therefore the number of ducks racing, was greater than the number of punters who had entered was kindly overlooked by my fellow committee members. Luckily, in the event, the winning ducks had actually been purchased, so no harm was done. If an un-bought duck had won, my gross innumeracy would have been uncovered: there would had to have been a steward's enquiry, leading inexorably to a public scandal and my inevitable impeachment.

During the past year I have also baked a lot of buns and made a lot of sandwiches for Go Games and prize-giving ceremonies, so I was quietly confident of re-election. To be honest, I have enjoyed being involved and was quite keen on another year's 'service'.

Sadly, I have persuaded no one to call me 'Mr Vice-Chairman', not even Younger Boy, who will sometimes refer to me as 'Sir Dad' on receipt of a small bribe. I had hoped the Beloved might occasionally address me by my title, if only facetiously, but no. Not once – not even in bed. At this year's meeting, pretty much the same people pitched up as last year, and the common mood was one of general satisfaction with the committee, so there were no major upsets. There was a bit of a reshuffle – the chairman and the secretary swapped positions. After the duck race accounting fiasco I thought it best not to run for treasurer and decided to stay where I was if the electorate was happy. My reappointment went through on the nod after a swift proposal and seconding by two of the women on the committee. They have been appreciative fans of the chutney in my sandwiches and the icing on my buns. I am thinking of aiming for Vice-Chairman for Life. I might pull it off if I can keep the cakes coming.

Singing Carols

Five thousand years ago, people we know very little about built a complex of megalithic structures in the Boyne Valley in County Meath, the primary edifice of which is Newgrange. It is aligned so that on the shortest day of the year the morning sun shines down the entrance passage way and lights up the inner chamber. These days this happens a few minutes after sunrise, but someone who is very good at sums has worked out that when it was built it would have occurred exactly at dawn.

Newgrange is mostly constructed of a feldspar-rich sedimentary stone called Greywacke, with a few slabs of brown sandstone. To build it would have required profound astronomical knowledge and astonishingly precise engineering. Of the rest of these people's culture nothing beside remains. We have no real knowledge of their daily life but we have an enduring monument to their beliefs – we just haven't really figured out what those beliefs were. All that is certain is that the winter solstice was vitally important, as it continues to be for some.

Every year at this time, hundreds gather before dawn at Newgrange to witness the event. Today a lucky few were in the chamber, waiting – they were mostly disappointed, as the morning was overcast and the light didn't penetrate to the inner sanctum until well after sun up. Which itself demonstrates a deep truth: never rely on the Irish weather. The only thing that that can be said with any certainty is that it will generally be bad. This is sometimes helpful. The Romans came here but didn't linger. They landed briefly on the coast of Leinster, named the place Hibernia – or Land of Winter – and left. They didn't stay long enough to observe the native defiance of the elements. This is a country where an all-pervading, saturating drizzle is called a 'grand soft day'.

The solstice is the year's midnight: the midpoint of *gamh*, which is the old Celtic name for the dead half of the year centring on winter. This ran from Samhain to Bealtaine, the first day of summer, or Mayday. The other, living half was called *samh*. But these are terms that have long since fallen into abeyance – nowadays the word Nollaig is more generally used. This can refer to the solstice itself or December and the festive season generally – Lá Nollaig being Christmas Day. The derivation of the word *nollaig* is not absolutely certain. The commonly accepted idea is that it comes from the Latin word *natalicia*, which means a birthday party – the obvious association being the birth of Jesus.

Another suggestion is that it comes from the Irish words *nua* and *lá* – 'new day'. To my mind, this pleasingly evokes the idea of the solstice and the lengthening of the days thereafter. A more obscure etymology suggests the word comes from an old term for 'elbow' or 'hinge' and this too suggests a turning point, a pivotal moment after which the sun grows steadily stronger.

The winter solstice marks the birth of the Sun in mythologies all over the northern hemisphere. The Persian sun god Mithras and the Semitic sun god Shamash were absorbed by the ancient Romans into their Saturnalian festival. This pagan week of debauchery and present-giving in December was absorbed in turn by the Christian Emperor Constantine and converted into Christmas. It seemed the perfect time of the year to celebrate the birth of 'the light of the world'. When Patrick came to Ireland he found a long established ritual acknowledgement of the year's rebirth – the time was ripe for appropriation, and gradually the solstice became inextricably combined with the birth of Christ. The date was fixed as the twenty-fifth of December in the fourth century, but with the switch from the old Julian calendar to the Gregorian one we use today, it became slightly out of sync with the solstice.

In the old Celtic solar-based year, Nollaig was of lesser importance than the two major 'fire' festivals on either side. It was overshadowed by Samhain – the end of summer – and Imbolc – the beginning of spring, but not any more. Christmas is now without doubt the most important festival of the year, at least in Younger Boy's opinion. He told me recently that it was 'the best day of his life' because he gets presents and chocolate figures of Santa. He said his favourite part of the day was coming into our bedroom in the morning and emptying his stocking in front of us. I wonder what his strongest memories will be when he is a grown-up? The Beloved and I are quite conscious of the fact that in arranging our own Christmas routines we are 'creating' memories for the boys to choose from when they are older. It will be their own subconscious selections that will define childhood Christmas for them as adults. My own most vivid memory is of a green scooter I got when I was probably about six or seven. It had yellow wheels and a metallic sheen to its pine-coloured paint. When I caught my first glimpse of it under the tree on Christmas morning, I stopped breathing for a second. It glinted from a dark recess of the room. The fairy lights reflected in the shiny green handle bars seemed to represent all the promise of all Christmases ever.

It is only in recent years that I have been able to recapture a sense of that seasonal magic, and this has mostly been here: in this small house, on this low hill, in sight of the yellow mountain. The Beloved and I are now usually the hosts for the day and we take pleasure in that. I think, for both of us, Christmas is, in part, a joyous occasion insofar as we can enable it to be so for others, whether they be friends, parents or, essentially, our two small sons. We have both, over the decades, witnessed, and indeed experienced, the stress and anxiety that often comes with the season: particularly the forced, false jollity; the almost desperate determination to 'be happy' – the rictus grin beneath the paper hat. Christmas Day can be a spiritual magnifying glass – it often amplifies deeper, unacknowledged emotions, sharpening conflicts or exaggerating darker moods. People gather at the coldest time of the year in search of human warmth. But beneath the shared quest for joy and goodwill lurk unfulfilled needs and suppressed unhappiness. Like Banquo's ghost at Macbeth's feast, they appear, unbidden and relentless. However, in the annual reassertion of the triumph of hope over experience, most of us doggedly believe that Christmas can and will be successful and fun if only everybody really wants it to be. This is despite the fact that the day often collapses under the massive weight of expectation. Or else it slips by anticlimactically in a numbing fog of alcohol and gluttony. Yet the memory of that green scooter still glows somewhere inside me, like an ember waiting to be blown into flame, and these days it doesn't take much to set it alight.

I am only a little ashamed to admit that I can find something to do the job in the flood of mawkish tosh on TV at this time of year. I was watching such a thing the other night. The boys were in bed and the Beloved had a few days' work in the city, for which we were both extremely grateful, so I was alone. I took the opportunity to wrap some presents and stuck on the telly. I channel hopped in search of a golden oldie like Jimmy Stewart in *It's a Wonderful Life* or Alastair Sim in *A Christmas Carol* – something in which soppiness is muted by black-and-white photography. All I found was a movie made ten years ago but already deemed 'a classic' by schedulers. Set in a luridly cosmopolitan west London, it follows the lives of various beautiful people as they negotiate the season in search of affection and happiness. 'Nollaig Hill' I suppose you could call it. It wasn't perfect, but I let it play as I wrestled with cheap wrapping paper and cheaper sticky tape, and I was inevitably drawn in by its tinselled saccharine. It is a vastly popular film and has been all over the airwaves this last while – the Beloved and I watched it together a few days later. I didn't mention I had seen it recently, as I was keen enough to wallow in it again. I pretended to myself that I was merely scrutinising the skill displayed by some actors I admire, but to be honest, I enjoyed it all. The sentimentality is just idealised shorthand for the human heart's desire. Over the last decade or so, I have gradually come to realise that I can now enjoy a truly happy Christmas because I am myself fundamentally happy. By which I think I mean that I finally understand love … actually.

When the Beloved and I first met, she wasn't keen on Christmas at all. This didn't surprise me too much – quite early on in our relationship I had discovered her profound impatience with anything which she considered to display cheap sentiment. On an early date I thought I would show her my sensitive side so I took her to the cinema to see *Les Parapluies de Cherbourg*. I thought this was a cunning ploy – a French musical all about *l'amour* would demonstrate that I was afraid of neither emotions nor subtitles. As the last song faded and the credits rolled she asked me what I had thought of it. The film starred a young and sensationally beautiful Catherine Deneuve, so I was selectively honest when I said I thought it was wonderful. Then I asked her what she had thought and she said, 'I wanted to slap them all.' As that Christmas approached, I thought better of taking her to see *Miracle on 34th Street*, so we just swapped presents over a good dinner in a fine restaurant and went our separate ways to meet familial obligations. This sense of obligation lingered for a year or two, until one December we realised that neither of us had to be anywhere in particular on Christmas Day. My mother was staying in England and my father was in Africa. The Beloved's father had sadly died that year and her mother and sister were spending Christmas in France. The two of us had professional commitments which required us both to stay in the city for the duration, so we were free to do as we liked and Dublin was our oyster.

We had no one to please except ourselves, so on Christmas morning we opened our stockings (I had insisted on these) and toasted each other in a glass or two of a decent pink champagne. Then we fortified ourselves with a full Irish breakfast and went for a walk around the quiet streets. It was a bright, frosty morning and the resonant bells of Christ Church and St Patrick's, the city's two medieval cathedrals, were tolling jubilantly. We strolled at our ease in the crisp air and then, feeling a little appropriate music would be nice, we wandered into Christ Church to hear the choir sing the last sequence of a Mozart mass. The ethereal harmonies seemed to linger in the fan vaulting of the ancient church's high stone roof. As the service was ending, we made our way down a side aisle and crossed the bottom of the nave to leave by the great door. Then we noticed the archbishop himself, seasonally greeting the congregation as they were filing out. I have never met a prelate so I suggested we hover for a moment to shake his hand. The Beloved said, 'I'd like to see you try,' and nodded at the long line backing up the central aisle. There was a solid phalanx of formidable women in formidable hats queuing to meet him. Their collective gaze was like a psychic force field daring me to push in. I backed down hastily and we left. We wondered how many diners that day would hear their hostess declare, 'As I was saying to the archbishop only this morning …' We also acknowledged that, regrettably, we would not be able to do the same.

We went back to the Beloved's city centre flat to swap presents over some foie gras and an unctuous wine from Sauternes, and the day continued in a similarly relaxed and indulgent manner. In the afternoon we went for a short stroll along the North Wall pier. A gentle breeze was blowing a light spray. The day was bright and the sea glittered. We could clearly see both ends of Dublin Bay – Howth Head to the north and Bray Head to the south. The salt air sharpened our appetites nicely, so we headed back into town for our dinner. Later that evening I was in the kitchen, having cleared away our starter of salmon and Meursault, when I heard the Beloved crying softly. I went to see what was up. It was her first Christmas without her father and she was missing him. I lingered in the doorway until she gathered herself and smiled at me. I returned to the task of preparing the main course. It wasn't long before I heard her hoot with laughter. This was prompted by the fact that from where she was sitting she saw the duck we were planning to eat fly past the kitchen door. I was serving the bird with a fine Saint Émilion, which I had been testing for general tastiness before carving. Roasting a duck produces an awful lot of fat: the large amount of grease in the roasting tray and the large amount of wine in me combined to produce a double lubrication. I plied the carving knife less dexterously than I had intended and accidentally managed to render the duck airborne again. It was a short flight. The duck landed with a slap and slithered to a halt by the fridge. I speared it with the carving fork, mopped the fat off the floor with the apron, steadied myself and served. It was delicious – slow roasted, moist and falling from the bone. I think we might have burnt the pudding later. My memory is a little blurred and, to be honest, I can't really recall, but if we did, I doubt we particularly cared. That was the only Christmas Day to date that we have spent entirely alone, and it was lovely.

In general, though, during those city Christmases in the early years of our relationship, the Beloved tended to downplay the significance of the day – especially the dinner itself. This central ritual, it seemed to me, oppressed her with its sense of social obligation and the associated pressure to celebrate 'the Christmas spirit'. One year she avoided it altogether by refusing all invitations and spending the day volunteering in a hostel for the homeless – which is the best demonstration of the Christmas spirit I can think of. However, she was never averse to a little seasonal music, and over those first years we developed a custom of our own. I would have been happy to attend a concert of cheesy Christmas number ones from the seventies, but my soul is a little more vulgar than hers, as I realised when our first Christmas loomed and she bought us tickets to hear Handel's *Messiah*. We went to the National Concert Hall to hear the Oratorio ... in its entirety. I liked bits of it and I was reminded of the remark someone once made of Wagner – that he had wonderful moments but terrible half hours. I admit to dozing off here and there till I was woken fully by the 'Hallelujah Chorus', which is of course fab. I claimed I had my eyes closed to concentrate on the composer's use of counterpoint, and I quickly realised that she was not going to stand any of my nonsense. I had also realised that she liked sophisticated concerts – for her part, she realised I preferred edited highlights.

The next year we found the perfect compromise when we went to a carol service at St Patrick's. When the church was built in the twelfth century, it was consecrated as a Roman Catholic place of worship, but since the Reformation it has been a Protestant cathedral and therefore heir to Anglican religious traditions. I have long been fond of English choral music. It is one of the great cultural gifts England has given to the world, along with Shakespeare, the BBC and pork pies. In my twenties, whenever I was on tour around the UK and was too far from London to get home on a Sunday, I would find the nearest gothic cathedral (with which England is amply endowed) and go and listen to sung evensong. It was a solitary but often sublime experience. If I was lucky, the late-afternoon sun would be shining through kaleidoscopic stained glass and multicoloured light would fall on the white vestments of the choir as their interwoven melodies caressed the cold carved stone. It could easily confirm the beauty and majesty of God – if you believed in him. But for me it was always a secular pleasure. The transcendence I felt was because of beauty created by mortal hand and eye in glorification of an idea in the human mind. This particular style of devotional singing is a musical tradition that has evolved over centuries, refining and perfecting the interplay of its three key elements: the choir, the organ and the acoustic of those massive churches. It is quintessentially site-specific music and I have no doubt that its composers over the ages have responded directly to the architecture. The analogy is obvious: the organ is the foundation; the basses and baritones form the supporting columns and walls; the tenors make the high, arching roofs; and the boy sopranos are the sky-piercing spires.

When I was a small boy, that first melancholy Christmas in England – missing my father, grandparents and extended Irish family – I remember seeing the Vienna Boys Choir on TV and yearning to be one of them. Their high, haunting harmonies took me somewhere 'other' and touched my heart. Boy sopranos still have that effect on me, especially at Christmas. Every time I hear the descant to the third verse of 'O Come All Ye Faithful' I choke up. I particularly adore the line 'sing in exultation'. When the combined voices leap up to harmonise joyously on the high notes of the word 'exultation', it is a perfect union of form and meaning. And when you hear this performed by a good choir, in a building the music was designed for, at the right time of the year – like St Patrick's Cathedral on the Sunday before Christmas – well, it has a glorious, heart-bursting beauty.

Apart from the flit through Christchurch to catch some of the Mozart mass (which was commissioned by a Catholic Austrian Emperor) and one visit to Dublin's Catholic cathedral to hear the Palestrina choir, our customary seasonal musical experience came to be a Protestant one – the Nine Lessons and Carols at St Patrick's. As baptised but non-practising Catholics, religious affiliation is irrelevant to us both. I do, however, have a vague half-remembered and possibly misapprehended childhood sense that Protestants harboured a slight sense of superiority or 'notions of upperosity' as my grandfather called it. He was fond of quoting a ditty to puncture any such pretensions: 'Away with the foreigners' church. / Lacking meaning or foothold in faith / For the stones upon which it is grounded / Are the bollocks of Henry the Eighth.' My grandfather had a tongue both rich and sweet and a faith both fierce and deep – five priests concelebrated his funeral mass. I know he was fond and proud of me but I suspect he would spin in his grave if he knew that the only church I chose to frequent was a Protestant one. Once I went so far as to become a 'Friend' of the cathedral in order to ensure getting into the extremely popular Christmas Eve gig. An additional perk was a summer invitation to the dean's garden party. I still regret not going – I might have got away with wearing a straw boater.

The Lessons and Carols varied from year to year, but the finale was always the same: Handel's version of 'Hark the Herald Angels Sing', with a rousing trumpet fanfare to accompany the choir and organ. This was the sole number where the congregation was allowed to join in. The only sectarian generalisation I would risk making is that Protestants take pride in singing very loudly, and for this last carol the challenge was always to drown out the brass. It was often achieved and we both enjoyed our annual church visit: for the fun of the final bellow and for the aesthetic qualities of the overall service. But I was never emotionally moved on a personal level until the last time we went, nearly ten years ago now. The choir sang a carol I had never heard before called 'Bethlehem Down'. The arrangement was exquisite but it was the words that caught me unawares and moved me almost to tears. They spoke of the newborn infant boy asleep on his mother's breast, peacefully ignorant of the violent, bloody death that was to be his future. One verse runs: 'When he is king they will clothe him in grave sheets / Myrrh for embalming and wood for a crown. / He that lies now in the white arms of Mary / Sleeping so lightly on Bethlehem Down.' When I heard the choir sing it, my eyes moistened. It evoked the very human tragedy of an ancient Galilean radical preacher whose agonising execution was witnessed by the mother who had suckled him as a baby. At the time that I first heard this beautiful carol, I was emotionally ripe and vulnerable. I was open and ready for new and profound things, not because I was about to become a believer in the divinity of Jesus of Nazareth, but because I was about to become a father.

That Christmas the boy was still a month from being born, but already I was full of the incipient anxiety of the new parent. The idea of a tender, helpless baby destined for crucifixion was devastating. The Beloved was touched and a little amused by the way the carol had moved me. She was more amused a few days later when, back down in the country, we were supermarket shopping in our nearest town. She caught me suppressing a sob as I pushed a laden trolley along a crowded aisle – the piped music was playing 'When a Child Is Born' by Johnny Mathis and it was too much for me. Ever since then she has entertained a sense of moral ascendancy when it comes to taste in Christmas music. However, this year I think a balance has been achieved. The night we decorated our tree, I observed her wearing a Santa hat, without a hint of irony, and singing along to a kitsch jazz version of 'Chestnuts Roasting on an Open Fire'. My long campaign to rehabilitate Christmas in her eyes is paying off.

Incidentally, I have since discovered that 'Bethlehem Down' was written as recently as 1927. It was the winning entry in a competition set by an English newspaper. The words were by a pissed poet and the music by a practising druid. Apparently they were regular drinking buddies and they wanted the prize money to finance a stupendous binge. I think the Jesus who turned water into wine would have approved.

We don't visit the great cathedrals any longer – when the boys are older, perhaps. These days our seasonal music consists of whatever the boys sing at their Christmas concert and a few carols at Got a Few Bob Friend's house. This is a tradition he has inaugurated over the past few years. For an hour or so on Christmas Eve, a gang gathers around the large fire in the large drawing room of his large Georgian house. He distributes lyric sheets and, led by his mother on the fiddle, we shout and roar our way through a dozen favourites. The children tend to run away and play elsewhere. They don't seem to like it much – this year Older Boy was especially embarrassed when I attempted the soprano part to 'The First Noel'. Our singing always begins a little raggedly – the introduction usually being, 'One, two, three, go!' But after four or five carols and four or five drinks, I think we make a tolerable sound.

There are a few other seasonal habits that have solidified into customs over our time here. I make the Christmas cake and mincemeat towards the end of November and at about the same time a neighbour over the hill makes excellent plum puddings. Each December we all expect a swap: we get a pudding in return for a jar of mincemeat and a jar of chutney. I feel that we get the better deal, but perhaps she thinks she does.

After I've made the cake, I put it away so that it can fully absorb the large amount of brandy I stick in. So I sort of forget about it until we put up our tree. Now it seems wrong to cut the cake until the ritual dressing of the tree is complete. The day after the tree goes up I cut holly and ivy from our hedge and drape it around the pictures in the sitting room. A month ago the holly was bright with berries, but they had all gone by the time it came to cut some. The birds must have been gorging themselves. I also fashion a wreath from the lower fronds of one of our leylandii trees, tie it with ribbon and hang it on the front door.

For some successive years running we ate a locally reared goose on Christmas Day. The first year we did this my mother was with us for the holiday and I brought her along when I went to fetch it. Older Boy was not quite a year old at the time so we strapped him into the back of the car and the three of us headed off over the hills. Male Farmer Friend knew of a woman who reared, killed and plucked geese herself. He ordered one for us and then gave me her name and rough directions to her farm. However, one country lane looks pretty much like another, so we soon got a bit lost, but my mother was enjoying the ride and I was enjoying the quest. Mum was looking contentedly at the rolling hills around us, and through her I felt a link with past Christmases at my grandmother's house and from there on back through the generations. My first born was gurgling in the back and, although he will never remember his first Christmas with his grandma, I will. I didn't quite know where we were but I was unconcerned. I felt content and complete. 'In the middle of the journey of my life I came to myself.'

As it happened I was on the right road. We were driving vaguely along when a little green post van emerged from a dark wood up ahead and came bouncing towards us. As we were squeezing past each other, windows were wound down to exchange greetings and I asked the postman if he knew where such a one lived. He said, 'Go straight on and you'll drive into the kitchen.' His directions were precise and we found the place. Our goose was waiting on the large oak table. The woman of the house insisted we had a quick cup of tea while I settled up. Her mother smiled and nodded at my mother from her chair beside the fire. Everyone admired the baby and we drove home. It was the first goose I remember eating and I won't forget it. There is not as much meat on a goose as there would be on a turkey of similar size and weight but it is extremely tasty. Like a duck, it produces a lot of fat. We always save this and keep it in the fridge for making perfect roast potatoes. The Beloved tells me that goose fat is also good for rheumatism. She remembers seeing it slapped on horses' legs when she was a child. I am a little suspicious – she often derives much merriment from the gullible enthusiasm I tend to show for anything I believe to be an obscure country custom.

That first Christmas goose came from about five miles away; the ones we ate over subsequent years were much more locally reared. The Farmer Friends used to rear a dozen or so each autumn – some for sale and some for themselves, family and friends. We would put our name on one and occasionally visit the flock to watch them grow. Geese are strong, noisy birds and make great night alarms – if disturbed, their combined gabble is quite deafening. The boys enjoyed the honking racket they made whenever we approached with a bucket of grain. The geese grew fatter as the days grew shorter, but each year, when the time came, we couldn't get the Farmer Friends to accept payment. The first one they insisted was in lieu of a present – we were swapping gifts at this stage. Another time they insisted the goose was in return for some tickets I had arranged for the Christmas show I was in at the time. I pointed out that the tickets were my allotment of complimentary ones and therefore hadn't cost me anything. They replied the goose hadn't cost them anything either. One year they rang to delay my popping in to collect our goose. They had been really busy and were a bit behind schedule. Male Farmer Friend apologised and said he would drop it down to us. I said I didn't want to put him to any trouble. He said, 'No bother, sure it's only a short walk down the hill!' and hung up. This worried me. A short walk, he had said. Was he going to lead a live goose down to us and leave it tied by a bit of stout twine to our door? I wouldn't put it past him. I had recently asked him about the correct way to kill a goose, and he had told me in detail. It involves using a broom handle to pin the head to the floor. You place your feet on either side and yank the large body upward, thereby breaking the thick neck. 'You should give it a go!' he had said – did he mean now? To my relief, the goose was plucked and drawn when it arrived. He was just amusing himself at my expense. I have given him a good few laughs over the years and not all of them on-stage.

That particular bird was eaten by my father and stepmother when they came for Christmas a couple of years ago. Stepmother, as a term, has rather negative associations and she has never liked it. Once we had a short holiday together in Provence and discovered that the French have a useful designation: *belle mère*. It is an umbrella term which can mean a stepmother, a grandmother or a mother-in-law. We tried it out for a while. It seemed to work quite well in situ, but it didn't stick outside France – she is, after all, a Jewish South African Buddhist, and a French title would have been gilding the lily.

The boys call her Bobba, which is the Yiddish word for granny, and she has taught me many other fine Yiddish words over the years. One comes to mind when I think of that winter visit. She and my dad don't make much of a fuss at Christmas – it really is a festival of the northern hemisphere – but they got into the mood well enough whilst they were here. I remember her wearing a set of illuminated reindeer antlers as she gnawed on a large, greasy goose wing with obvious relish, and I thought, 'There she is … revelling in the schmaltz in both senses of the word.' In general usage the word *schmaltz*, of course, means excessive sentimentality, but the original Yiddish meaning refers to the oily fat that drips from a roasted fowl. After dinner Dad and I sang a few Irish rebel songs – me on guitar and him supplying verses in Irish. They had left Capetown in the height of the South African summer to arrive in the iciest winter we had known for years. The cold gave them both a bit of a fright. I doubt they will spend Christmas with us again.

My mum comes over from England when she can and stays with us for a week or two. And the Beloved's mother joins us for Christmas dinner most years. Last year she suggested poetry readings after dinner. Older Boy kicked us off with a rhyme about a reluctant turkey which he had learnt at school, then I reached down a volume of Patrick Kavanagh's verse and we all read personal favourites. I recited the ones about the poet's own rural childhood Christmases. This was for the benefit of the boys – I wondered if they would hear any echoes of their own rural Christmas – but they were too absorbed in their new gaming console to pay any attention. We haven't had goose for the last two years, and may not for some years to come. The boys prefer turkey and, inevitably, for the foreseeable future the customs which harden into tradition will be the ones that they deem important.

Filling Stockings

It used to be traditional in the countryside to make a ritual acknowledgement of the travellers to Bethlehem. On Christmas Eve the door was left unbarred, bread was left out and the table was set for three. We set the table for Santa. We leave him mince pies and milk. This year Younger Boy suggested we leave him a fiver as well, but changed his mind when it was suggested the fiver might come from his own piggy bank. He and his brother practically galloped up the stairs to bed – Christmas Eve is the one night of the year when they are actually eager to go to sleep. This gives us a little breathing space to relax alone, but it has its disadvantages too. The earlier they go to sleep, the earlier they rise, naturally, but it also increases the chances of them waking in the middle of the night and unmasking Santa. Older Boy woke unexpectedly on Christmas Eve last year and it was an uncomfortably close thing. They had hung their stockings by their beds as usual and, just before turning in myself, I had taken them downstairs to fill – as usual. I had just climbed to the top of the stairs with the open door to their room on my left when I saw Older Boy, just awake, looking puzzled. He spotted me and said, 'Dad! Our stockings are gone!' I froze and thought quicker than I have ever done before or since. I realised he hadn't spotted the bulging stockings in my hand – they were just out of his line of sight behind the door frame – and I said, 'Ah … yes … I noticed that there was a hole in the bottom of your one and I thought it would be a great shame if any presents fell out after Santa had filled them so I took the stockings downstairs to stitch them up before he came!' He said, 'Thanks, Dad,' and turned over to go back to sleep. I tiptoed back downstairs and had a whiskey to steady my nerves. I waited till he was deeply unconscious before slipping the stockings back into place and diving into bed.

Like most parents, I want to sustain the belief in Santa for as long as possible. In the minds of small children he is tangible evidence of a glorious and generous universe. This was emphasised for me when the boys were six and four respectively. One morning in mid-December of that year I found myself staring intently at a middle-aged man in a penguin suit. He was gesticulating wildly and squeaking at me. Under his flapping beak I could see his unshaven chin. The skin beneath the stubble had the blotchy, waxy look of someone recovering from a night on the tiles, and his beer-soaked breath was coming at me in waves. This was not pleasant as I was feeling a little nauseous myself. I was hung-over and sweating tropically under a padded belly and a heavy Santa suit. The spirit gum attaching my luxurious white beard was dissolving slightly and I was concentrating hard on speaking loudly without moving my lips too much in case the thing fell off, thereby destroying the illusions of the two hundred five-year-olds who were our delighted audience.

I was touring with a show called *Santa and the Penguins*. I was playing the main man and it was a big responsibility. I had just played Rochester in an adaptation of *Jane Eyre* for the same company and was all set to play Scrooge in *A Christmas Carol* when the production annoyingly fell through. It was a disappointment creatively and financially, and in recompense the company offered me the part of Santa in a show they were doing for toddlers. I needed the money so I begrudgingly accepted – feeling that children's theatre would be totally dull and not very rewarding. I was utterly wrong and have rarely enjoyed a role as much. The costume was top quality and so were the excellent white wig and beard. With padding and a touch of make-up I looked pretty convincing. The director had told me that under no circumstances was I to appear wandering around the theatre between shows wearing anything less than the full monty. Otherwise I must be completely out of costume and make-up – no lounging around between shows in half-dress. He was quite adamant that the children must believe that I was the real thing and that I had come to tell them all about my concerns up at the North Pole, and therefore it was vital that we never undermined the pretence. (We will conveniently ignore the fact that the penguins in the show could only have been resident in the southern hemisphere.)

I thought he was being a bit precious ... until the first show. The wave of adulation that hit me when I came on was astonishing. It was like being a rock star for tots. But however deafening their welcome, quietening them was easy: they always did my bidding instantly – the suit was omnipotent. When I calmed them down I would chat to them for a minute or so before we began the story. I could see their expressions as they called out to me. Ecstasy is the only word.

Sometimes a teacher would ask me to name-check a birthday boy or girl and I was always delighted to oblige. On one occasion I was asked to come and meet a little boy who was gravely sick. I went to find him after the show. His little friends very gently nudged him towards me. He was extremely pale, extremely thin and extremely weak. His mother had prompted me earlier, so I greeted him by name and told him I had received his letter and would definitely be bringing the remote-controlled red racing car he had asked for. He absolutely beamed at me. I felt a warm prickle behind my eyes and a hot surge in my chest. I managed to turn it into a chortle; then I said something about getting on with overseeing the elves and went backstage. Everybody should play Santa once. It was a joyous experience. Sadly the boys couldn't really come to see the show. Which was a shame as it is the only theatre I've done since their births which has been age appropriate. We felt that they probably would have recognised me, which may have subverted their own belief a little, and they may have shouted out, 'That's my dad!' which wouldn't have been ideal. I would have liked to see them jubilantly yelling with the other children, but it doesn't really matter – after all, I get to see their faces on Christmas morning.

The day began quite early this year. They were in to us with their stockings at six thirty. When I woke up I heard an eager whispered debate coming from their room next door. Older Boy was arguing for waiting an hour before coming into our room; his brother was saying that he couldn't wait and was going to empty his stocking immediately. I take great pleasure in choosing the contents of their stockings (basically, I buy things I used to love getting in mine when I was small) and I didn't want to miss anything, so I called them in. The Beloved groaned softly and then they were upon us. Half an hour later the bed was littered with tissue paper and the tinfoil wrapping from chocolate coins, along with knights in armour, pencils, rubbers, bandit masks, balloons, marbles and fart-whistles. They were both pretty pleased with their haul and both bemused by Santa's continuing habit of leaving a tangerine in the toe of each stocking, as they never ate them – as is now customary, they gave them to us.

I went downstairs to get tea and juice and to light the stove. While I was waiting for the kettle to boil I went out to the piggery for more wood. The sun was up but hadn't been for long. Sliabh Buí was silvery blue and the pearly early-morning light was glistening on the frosted fields. Away to the west, the last of the night sky was dwindling into indigo clouds on the dark horizon. The iron bolt on the piggery door had a thin coating of ice and I had to give it a belt before I could pull it open. It was a perfect Christmas morning. The boys would have adored snow, but after two consecutive frozen winters a couple of years ago, I am less keen on a white Christmas than I once was. Not that I would want it to be too fine a morning: warm weather feels a little fraudulent on Christmas Day – as though you are celebrating the occasion under false pretences. A little bite in the air and a slight crunch underfoot is perfect. It is also good luck: cold, icy weather on Christmas Eve was thought to signify a mild spring ahead and an absence of illness in the coming year. We seemed to have had exactly that. The water bucket we had left out for the reindeer was iced over and their carrots were slightly frozen. If I had been out for daybreak I would have seen 'the wonder of a Christmas townland, / the winking glitter of a frosty dawn', as Kavanagh described his childhood Christmases. I kicked over the reindeers' bucket and half-chewed a few of the carrots. Then I scraped a greeting to the boys on the frosted windscreen of the car, signed it 'Santa Claus' and went back inside.

The rest of the day proceeded as I imagine it to have done in countless other households.

That night, I was sitting on the bench looking at the stars. I was drinking better wine than we have been used to recently and smoking an excellent Cuban cigar which had been a gift from a thoughtful friend. The day had been a peaceful, smiling pleasure but I had an obscure sense of anxiety. I suppose there is always an inevitable sense of anti-climax on Christmas night. In years gone past I sometimes had a sense of hollowness after all the hoop-la leading up to the big event. I have memories of being disappointed with the day after it was over, of having a notion that I had missed something which everyone else had grasped. Was this feeling of deflation a subconscious association with all those old Christmases? I re-examined the day as it had just unfolded. Maybe there hadn't been quite that giddy sense of delirium the boys had exhibited when they were a little younger, but their day had been an undoubtedly happy one. For our part, the Beloved and I had thoroughly enjoyed ourselves. We had shared a cheerful drink with friends in the middle of the day, followed by a brief but pleasant walk in the fields around us. We had planned ahead and dinner was delicious and stress free. The Beloved's aunt, who had joined us in mid-afternoon, was delighted to be with us and had brought generous gifts and decent wine. She too had partaken of, and contributed to, the general mood of contented ease.

I took another sip of the aunt's wine and another puff of the friend's cigar and contemplated the deep, black night pierced with a billion stars. It was vast, austere and glittering. I must have the memory of a goldfish because I am regularly surprised by the brilliant immensity of the clear night sky. I keep forgetting how astonishing it is and find myself in awe of it all over again. Looking at it is always helpful whenever I have an unquiet mind but am not quite sure why. It helps clarify what I think and feel. As I ponder the cliché of my insignificance in the infinity of the universe – as millions have done before – the colossal, cold beauty of our galaxy clears my head and makes me cool and calm. My conscious thoughts are silenced by the vastness, and the small, bothersome details in the back of my mind flicker into life, then become self-aware and self-absorbed. They proclaim their primacy and acquire my attention.

I was watching the shreds of a cloud briefly veil the face of the moon. It seemed more intently white after the clouds had cleared. I stared blankly at it and, sure enough, the fretful, niggling things floated to the surface. At first I recognised the familiar background radiation of sustained low-level panic about money, and where the hell we were going to find any. I quickly suppressed that repetitive, boring thought and listened to a new, more interesting anxiety that was bubbling up. That morning Older Boy had been a bit blasé when he and his brother discovered the message from Santa in the ice on the car windscreen, and he had also seemed distinctly unimpressed when they both saw the evidence of the reindeers' hasty eating and drinking. Then I remembered that earlier this year he had expressed profound doubts about the existence of the Easter Bunny, and throughout this particular day gone by, his demeanour had been more circumspect than usual.

Maybe this was just because he is getting older and therefore feeling a need to assert his maturity compared to his brother, or maybe … it was because he was beginning to lose his faith in Santa! I was quite upset by this thought. The more I considered the possibility, the more anxious I became. But why should it matter? He had still been delighted with his Christmas Day and, besides, he is a smart boy – if he was beginning to smell a rat, then surely this was an affirmation of his growing insight and perception. Oughtn't I to be proud of his developing intelligence and awareness, instead of wanting desperately to perpetuate an innocent but infantile fairy tale? I felt a small but forceful wave of sadness, not for the boy, but for me. I realised that, of course, the uneasy melancholy I felt was because of memories of my own childhood disillusionment with the Christmas myth.

The first Christmas I didn't believe any more was when I left Ireland. A seed of doubt had been sown the year before when my mother had taken my sister and me to visit the Santa in Switzer's – a famous store in Dublin, now long gone. It was a big deal at the time: their Christmas grotto was a celebrated institution and you got good presents. My sister and I were wearing matching balaclavas which my granny had knitted for us, and these disguised our distinct boy and girl haircuts. We each sat on a knee and Santa chatted to my sister first; then he turned to me and said, 'You're a lovely little girl. What would you like for Christmas?' This faux pas infuriated me and seriously challenged my belief in his omniscience. He correctly interpreted my scowl and did his best to recover, but the first crack had been made, and by the time I was in England the following December, the doubt was nearly unassailable. I say 'nearly' because I do remember lying in that unfamiliar bedroom on Christmas Eve and making a clear decision to believe one last time, despite that deep down I knew the bleak truth. I did my best and I almost persuaded myself, but not quite. Apart from other considerations, in this strange country they talked of 'Father Christmas' and the other kids in my new school laughed condescendingly when I mentioned Santa Claus. It was all wrong.

Sitting on the bench, looking at the stars, I realised that I have, over the years, associated my loss of belief in Santa with the break-up of my family and my separation from my father. The fading of that gratifying fantasy is connected, in my emotional memory, with a time of deep unhappiness in my younger life. I look at Older Boy now and I often see myself. As I watch him grow I am sometimes ambushed by the immediacy of long-forgotten feelings from when I was his age, and I have to steady myself and say, 'That was then but this is now.'

I thought of my father and how a few days previously he had texted me to arrange a time to Skype – Grandad and Bobba wished to speak to the boys before leaving the city for the holidays. In the event, it was a frustrating call. I had managed to get the boys to sit at the computer at the designated time, which is not always achievable. The audio link was fine, we could all hear each other perfectly, but we couldn't get the video signal working properly – they could see us but we couldn't see them. I could hear Dad tapping furiously at his keyboard throughout, trying to get the camera at his end to function – our end seemed to be fine. The boys pulled faces and chatted amusingly enough, but after a while they grew bored with the one-sided conversation. They chorused, 'Happy Christmas!' and strolled off. I tried to have a chat, but Dad persistently tapped away, now cursing a lot, as he kept trying to fix the visual transmission. I suggested he give up and just talk. He told me he had dressed as Santa Claus and was determined that the boys should see him. I pictured his brightly lit study overlooking the sea: the hot summer sun blazing through windows and him sweating and swearing in a red suit and white beard. We never made the connection. He was very disappointed and I was too. I would have liked to see my father dressed as Santa for the entertainment of my son. Quite apart from the amusement denied by his recalcitrant computer, I was touched by his desire to be present in some way and sad that he had been thwarted. With each passing festival his absences are accruing and he is becoming more of an abstraction in his grandsons' minds.

On St Stephen's Day, as has now become usual, we went to see the hunt off. Every year this meet is held in the larger of the two neighbouring villages and many riders and spectators gather. It is a popular day to go out. For a lot of people, this is the only occasion during the season when they beg, borrow or hire a mount and saddle up – so the 'field' is usually the biggest of the year – and after the over-indulgence of the day before, I find it pleasant to contemplate people gathering for some strenuous exercise without having to engage in any myself. This year's particularly large crowd obviously thought the same, but the village has four pubs facing onto the spacious central square so there was room and nourishment enough for all. A wide range of horses were gathered: sleek hunters with braided manes; pot-bellied ponies with unkempt forelocks; and hairy-legged workhorses with tangled tails.

There was an accompanying variety of outfits on display. The master and the huntsmen are the only ones to wear 'pinks', the scarlet jackets associated with foxhunting, and they looked dazzling in the crisp light. Others looked equally elegant, if more muted, in black or dark blue hunting coats – and besides these, there was a range of fleeces, anoraks and hoodies with the odd bit of denim in the mix. The clothing of the rider didn't always match the look of the horse. There was one young girl who sparkled astride a scruffy, evil-looking nag. The horse could have been painted by Breughel, but the girl seemed to glimmer. I looked closely and realised she was actually glimmering – she had some sort of glitter sewn in to the lapels and sleeves of her otherwise traditional and snugly tailored tunic. I nudged the Beloved and said, 'That girl has sequins on her jacket!' I thought she might think it improper in some way. It didn't bother her as much as my calling her tunic a 'jacket'. It was, of course, a 'coat'.

The Beloved grew up in a racing stable. Her father and grandfather trained thoroughbred horses – indeed, the granddad was also a champion jockey. Both their names are inscribed in gold lettering in the Turf Club rooms at the Curragh Racecourse – along with other winners of classic Irish races. As a small girl,

she and her sister would saddle up their ponies and disappear for the day, happily exploring the wide grasslands of Kildare. She is an excellent horsewoman and a fine judge of horseflesh. She has often taught me the difference between a piebald and a skewbald, and I have often asked her to repeat it just one more time. It is decades since horses were a key part of her life, but she loves being around them and I love that she loves that. I only learnt to ride a year or two ago. She gave me a present of lessons and I really took to it. We cannot afford horses of our own, but once in a while we are able to borrow a pair of mounts and we head up the hill and into the forest. Riding out together through the silent pines, jumping fallen trees and cantering just beneath the branches, is a shared, intense pleasure. I have been an apt pupil but I still have much to learn, not least the equine jargon. I try to remember the correct terminology (of which there is much) and I do my best not to let her down by demonstrating that I am an ignorant gobshite.

Hunting has a separate language all its own and I don't speak it, despite occasional amateurish efforts. One Stephen's Day past I was admiring the colourful scene with its sights, smells and sounds. I mentioned the excited barking of the dogs. The other people within earshot didn't bat an eyelid, being far too well-mannered to say anything that might make me feel awkward, but I knew I had blundered. The Beloved explained: dogs are called hounds and they never bark, they babble – unless they are on the scent of a fox, in which case they 'give voice'. Nowadays I am up to speed with the hound/dog issue, but I was regretting my coat/jacket faux pas. I took a long pull on my pint – it was cool, creamy and reassuring. An opportunity to reassert my *savoir faire* arose. I heard an approaching canine clamour and casually said,

'Here come the hounds and the huntsman.' The Beloved nodded approvingly. Then I cocked it up by pointing out Got a Few Bob Friend on 'the brown horse'. It was not, of course, a brown horse – it was a 'bay'. Whatever his mount was called, he looked sharp: his 'coat' was natty and his 'bay' was glossy. It was the best he looked all day. He told me later that he fell off. He wasn't hurt in any way – he landed on his feet, literally, and then hopped back on – he is a good horseman. However, he got thoroughly mud spattered, then and later, as it turned out to be a pretty brisk day. Astonishingly, the hunt actually spotted a fox and even pursued it for a while until it disappeared. This caused great excitement and much eager galloping across very mucky fields. Disorder ensued. Apparently in the tumult and thrill of the chase, someone had giddily overtaken the master. This is not done, unless he (or she) specifically gives permission by saying, 'Kick on.' If the master says, 'Hold hard,' you stop. No questions asked. If you hear 'Hold hard' repeated, it means 'Fucking stop!'

Before the hunt set off, a certain ritual had to be observed. People emerged from one of the pubs carrying trays of hot ports and hot whiskies, which they distributed amongst the riders as a 'stirrup cup'. I saw the master sink two in quick succession, as befits his rank; then he nodded to the huntsman, who called the hounds to heel, and off they headed – down out of the village square and on up the rising road opposite. With a sedate rumble of many hooves, the horses followed, and we watched until they disappeared over the brow of the hill and on into open country. The Beloved was not going out with them this year as her aunt was with us, so we headed home – maybe next year.

The first time we attended this meet, the Beloved was heavily pregnant with our firstborn and she needed constant nourishment.

As the crowd gradually dispersed after the hunt had left, we drifted with it and found ourselves inexorably nudged into one of the pubs, not against our will. We had just settled into a nook by the fire, with our drinks and her snacks, when some lads with painted faces and battered straw headgear barrelled into the bar and started singing tunelessly. It was my first encounter with the Wren Boys. They belted out a short rhyming chant about St Stephen and the 'wran' (as they pronounced it), collected a few coins in a roughly fashioned box wreathed in holly and squeezed up to the bar where freshly pulled pints awaited. They sank these with practised speed, rattled the box for more donations and headed back outside – presumably into the neighbouring pub for more cash and more drink. I only heard a snatch of their song and was completely unaware of what I had witnessed but I asked around and was soon filled in. There was much discussion and some disagreement concerning the correct wording and length of the song, but an essential version would be:

The wran, the wran, the king of all birds,
On St Stephen's Day was caught in the furze,
And though he is little, his family is great,
So rise up, your honour, and give us a trate.
Sing holly, sing ivy
Sing ivy, sing holly.
A drop just to drink it would drown melancholy.

The Wren Boys were collecting money to 'bury the wren'. These days this is a euphemism for 'having a piss up' and maybe it was always so. Hunting the wren is an ancient custom whose origins are now obscure, but remnants survive in the other remaining Celtic pockets of Europe – traces of the ceremony linger in Wales, Brittany and Cornwall.

In Irish folklore the wren was the cleverest of birds. According to the old tale, the birds held a parliament to decide who should be their king. It was suggested that whoever could fly the highest should take the honour and, not surprisingly, that proved to be the eagle – a suitably royal bird. But as the eagle soared to its limit and began to tire, the crafty wren emerged from the eagle's tail feathers and flew higher, thereby winning the crown. Like a lot of other Irish stories, it demonstrates how stealth and cunning can conquer power. It also seems oddly apposite that, having established the wren's superiority, the people should then want to kill it. Perhaps this is because, like much else in Irish mythology, there is a fearful touch of the supernatural about the wren. The Irish word for the bird is *dreoilín* and this has its roots in 'druid's bird'. It was thought to have acted as a messenger between this world and the next. Scholars have suggested that within the tradition of hunting and killing the wren we can see the ghost of an ancient ritual sacrifice of a sacred symbol, and that the burying of the wren once represented the burial of the sun on the shortest day.

With the Christianisation of the island, the custom gradually moved to the first day after the birth of Jesus, and this was partly because of the wren's association with Stephen, the first Christian martyr. According to the old story Saint Stephen was hiding in a holly bush when a wren alerted his pursuers, thus giving him up to death and glory. Legend also has it that the position of Irish soldiers waiting to ambush Norman invaders was given away by wrens beating their wings on their shields. So the cunning king of the birds acquired an additional reputation as a traitor. Maybe the annual ritual hunting of this tiny bird gratified a sublimated desire for vengeance; and I can certainly empathise with a powerless peasantry gleefully, if only symbolically, settling scores with those that have betrayed, abused or cheated them. There are certain producers I would like to hunt and bury.

A few days after Christmas, the Beloved heard Older Boy proudly tell a friend that he and his brother had found a personal message from Santa himself written on the car window. I think his faith is still firm – we might get one more year.

FESTIVE MINCEMEAT

Ingredients

500g apples (I use our Russets), cored and chopped
225g shredded suet
1kg raisins, sultanas, currants, or whatever you like
350g soft dark brown sugar
zest and juice of 2 oranges
zest and juice of 2 lemons
50g chopped almonds
4 tsp mixed ground spice
½ tsp ground cinnamon
grated nutmeg to taste
6 tbsp brandy

Method

Put all the ingredients, except the brandy, in a large, oven-proof bowl and stir thoroughly. Cover the bowl and leave in a cool place overnight to allow the flavours to mingle.

The next day, pre-heat the oven to 120°C/250°F/gas mark ½. Cover the bowl loosely with foil and put in the oven for three hours.

Take the bowl out of the oven and don't panic about the mincemeat's appearance at this stage. It will look like it is drowning in fat but that is how it is supposed to be. As it cools, stir the mixture from time to time and the fat will mix evenly through the other ingredients. Stir in the brandy when the mincemeat is cold. Pack in sterilised dry jars and seal. It will keep in a cool dark place indefinitely.

This recipe makes almost 3kg so it usually does us for two holiday seasons. That is, if we haven't had to give it away as emergency Christmas presents!

Shearing Sheep

Bloody January is here again. Sliabh Buí is invisible under a wet blanket of fog and the sky in all directions is an extended, coagulated mass of dull, grey cloud. It has rained steadily for days now. The lower half of our field is a sodden, squelching marsh. The hard core and rough gravel in front of the house are sinking in the rising mud. The kitchen garden is like a disused cemetery. The planks I hammered into the stony soil to define and separate the beds are rotting and covered in moss. The abandoned chicken run looks especially desolate. This morning I dug up potatoes for dinner – there are still some scattered in the wet ground but they are scabby and slug infested. The remaining carrots are stunted. The leeks are weed-strangled dwarfs and the last few turnips are woody and inedible. I still haven't planted any garlic, there doesn't appear to be any work on the horizon and there are next to no logs left in the piggery. Everything is dreary and mouldering. Why the hell did I move here?

Sometimes the city is very alluring so this year we spent New Year's Eve there. Friends of ours who have hosted the night for many years realised that this would be the twentieth year since they threw their first party, and they wanted to make it a big one. They made a particular point of inviting out-of-town, and indeed out-of-country, friends. We have gradually lost touch with our old urban life over the last decade and we decided to go. Whatever one might feel about the required cheerfulness of Christmas, I have always been bemused by New Year's Eve and the need to celebrate the precise instant when the clock strikes twelve on that particular night. It seems an utterly arbitrary moment – well, it actually is an utterly arbitrary moment. I suppose the notion of a 'new year' is another echo of the solstice, but at this point in the festive season, I have had quite enough of that, thanks.

In Ireland, the first day of January was legally designated as New Year's Day as recently as 1751. Prior to that, the legal year began on 25 March, and for country people, the working year commenced with the beginning of spring, which was the first day of February. Yet on the night of 31 December, I still feel an odd obligation to 'see the new year in', and the Beloved and I still feel we ought to wish each other a happy new year at midnight – or first thing next morning if we haven't been bothered to stay up for it. I used to care a little more when I was younger. I remember one occasion when I was twenty and had missed the lift that was meant to take me from the pub to a party. I heard the bells ring out as I was walking alone along an empty street and felt abandoned and forgotten: I had missed it! Then I asked myself what it was I had actually missed apart from a bit of ritual nonsense. I found the house where the party was in full swing, someone thrust a drink at me and I was grand.

In general, New Year's Eve is a night that I can happily let slip by without much fanfare, although we did have a good one down here a few years ago at Got a Few Bob Friend's house. His wife is from a village near Edinburgh and she organised us into lines and pairs and drilled us in Scottish country dancing. We launched ourselves into it with little skill but much enthusiasm. It was completely knackering and great fun. At one point I went to check on the boys. The four of us were having a sleepover, staying in the converted basement directly below the room where the adults were cavorting with light hearts and heavy feet. Younger Boy was wide awake, looking startled and staring at the ceiling. He bewilderedly said, 'Dad! What are you all doing up there?' Most other New Year's Eves I only vaguely remember. But this year's midnight was notable because it was the first time our sons were still up with us to hear the bells toll.

We drove up to town in the afternoon to visit a place called something like Winter Wonderland. We had promised to take the boys ice skating, which they had been very keen to try. They were a little disappointed that they weren't able to zoom around instantly. They were convinced that they knew how to skate because they had seen Tom and Jerry do it and it looked easy. Thankfully there were aids available – little plastic ice-bound Zimmer frames in the shape of penguins – and they clung to these and managed to slither–slide around the rink. Older Boy was astounded at how slippery the ice was and kept warning me to be careful. He didn't feel the need to warn his mother, as she was gliding around effortlessly. I haven't skated since I was about fourteen and I had forgotten how uncomfortable hired skates can be. I hobbled and slid, and even glided briefly. It wasn't elegant but I remained upright. Afterwards we had chips and went on a Ferris wheel. At the top of each circuit we could see right across the city to the setting sun. The sky was pink and red to the west and navy blue to the east. With each descent we plunged back into the coloured fairground lights and the gathering dusk.

We were to stay the night with the friends who were throwing the party, and when we arrived the boys headed straight to the room which had been set aside for children. It had a huge television screen with game consoles and stacked DVDs ready to go and a large store of fizzy drinks, crisps and sweets. There was also the promise of pizza and more chips later, so they settled themselves in and happily set about creating a sordid mess. They achieved this admirably, as we discovered when we cleared up the following day. But it has to be said that the adults also made a significant contribution to trashing the place. Everybody had brought something to the party – canapés, starters, wine, cheese or desserts. Our contribution was a raised game pie and jars of chutney. An American guest supplied the decorations. She wanted to recreate the mood of those glitzy roaring-twenties New Year's Eve parties of her native New York and had therefore brought feathery tinselled hats, horns, whistles and streamers – lots and lots of streamers. The following morning it looked like an entire Fifth Avenue ticker-tape parade had passed through.

The adults' dinner was a feast. Our hosts supplied the centrepiece of a whole pig which they had slow roasted for a day and we sat in conclave over it for an hour. The large plate of crackling was passed between tables ceremoniously, like a chalice. Ireland doesn't have many traditions connected with this night but eating a big dinner is one of them. It was known as the Night of the Big Portion and it was believed that eating a large supper that evening would ensure abundance in the coming year. It was also customary in parts of the country to throw a cake at the door. This bizarre charm was designed to prevent hunger. A loaf or a barmbrack was smashed against the door of the house – from the outside: no point in making a mess indoors – and a rhyme was recited warning famine to retire for the coming twelve months. If there is any lingering power in honouring these old invocations, we should have an extremely prosperous year ahead of us. I mean the one about the big dinner, not the one about throwing the cake. I don't remember anyone flinging food around – mind you, judging by the state of the party rooms the next day, that may well have happened.

When midnight approached, the children were dragged from the TV room to join the countdown. Our two looked a bit puzzled by the noisy whirl of the roomful of adults shouting in silly hats. I was reminded of my own childhood bemusement at the absurdity of grownups having 'fun'. We gave them a hat and a trumpet each. They glanced sceptically at one another and then shrugged and joined in. I wonder if they think there is some mysterious, exotic pleasure to be had in such celebrations – something they will only fully understand when they are older. And I wonder if they will be very disappointed when they realise that there is nothing particularly special to be discovered and that the whole event is just a tiny bit pointless. Or maybe they know this already, because as soon as the New Year's hugging and kissing stopped, they snuck back to the telly. I followed them, ostensibly to ensure they weren't watching anything inappropriate, but actually to have a little lie down. I squeezed between them on the couch, watched five minutes of *Star Wars* and then snoozed for half an hour. By the time I rejoined the party the dancing had begun. There is very little I like more than twirling girls around on the dance floor – especially the Beloved. We dance well together and, over many years' practise, we have perfected some pretty good moves, which we proceeded to demonstrate. She is a generous woman and is quite happy to loan me out, so I was passed around for an hour or two whilst various women had a good go of me. I went to bed at about four feeling quite tired, slightly drunk and only a little bit used.

The next day I dispensed the raised game pie and chutney. It was a thundering success, which was a relief, because it would have meant a prodigious waste of time and effort if it hadn't been. The recipe suggested venison but any game will do and I had used a pheasant, a rabbit and some pigeons, all from the freezer. I stripped and chopped the meat and marinated it overnight in port and brandy. Meanwhile, I made a stock using the bones from the game along with onion, carrots and celery. The hot-water pastry was easier than I expected: you melt a lot of lard in boiling water, stir it into spiced and lightly sugared flour and then it comes together quite quickly. I gave it a little knead, rolled it out roughly and lined a large cake tin with it – squeezing it evenly up the sides. The pastry was then lined with streaky rashers to hold in the juices from the filling. I minced some lean pork with pork fat and added chopped parsley and more spices. Then I alternated this with layers of the game until I had filled the pie. The lid went on and the whole thing was baked for about an hour and a half. When it had cooled, I added gelatine to the warmed-up stock and poured it into the pie via the steam hole I had left in the pastry lid. Then I left it alone to set. The Beloved's version of this lengthy procedure would no doubt add some extra colourful details, such as collapsing pastry lids, or jellied stock leaking all over the kitchen floor, or florid oaths and lamentations, but her conclusion would be the same as mine: the completed pie looked magnificent and tasted great. It was my first attempt at such a thing and it was a triumph, but there is room for improvement. I will try it again next winter. I think I may have inaugurated a New Year's tradition. The Beloved just can't wait.

As we drove home on New Year's night we were recounting aspects of the party and she asked me what I had thought of the speeches. I recalled no speeches! If there had been speeches, I probably would have made one. It transpired that these had taken place whilst I was kipping in the TV room, so she filled me in. Not many people had spoken and, as it happened, those that did were the ones living abroad. Part of the Irish diaspora, they all acknowledged the importance of coming home. They talked about attending this same party year after year: seeing the same familiar faces, comparing notes and sharing references. They talked of friendships solidifying into family; customs deepening into traditions; and this annual event being a consistently reliable source of fun, comfort and joy. They also talked of the unwaveringly high quality of eating, drinking, singing and dancing to be had in that house. Then the room erupted into clapping and cheering for our hosts. The Beloved's eyes glistened as she told me all this. I suppose her emotion could have been because she was struggling with a hangover; and, after all, the speeches had been made by people who were full of drink – which can reduce you

to declaring unswerving loyalty and eternal friendship to a shifty barman from Kinnegad, as I know; but nonetheless, the room had obviously been awash with gratitude, generosity and love. It sounded to me like a perfect expression of what celebrating New Year's Eve could and should be. And I'd missed it.

On our return to the acre, everything was still muddy and dripping. But at least the mist was beginning to clear. This morning I looked over the road and across the damp fields. Beyond, in the middle distance, I could see a stand of trees etched against the remnants of a fog bank. Further away on the horizon Sliabh Buí loomed above the mist, reasserting itself in gradations of grey, blue and silver. There was a short respite from rain, so I got out the chainsaw and headed off in search of some fallen boughs which the Farmer Friends had alerted me to. For mostly economic reasons, I do my best to avoid buying fuel for the stove. It has been burning all day and every day for nearly five months now and that is a lot of wood; we could quite easily have spent the guts of a grand, if not more. But there is something else motivating me in keeping the home fire burning, something other than the depressing hint of defeat I feel if the woodpile is low. When I reach for the chainsaw, there is a symbolic dimension to the action as well as an actual one. Keeping the piggery full of wood, apart from being practically useful – or rather, because it is practically useful – has come to represent achievement as a 'provider'. When this winter is finished, and we have a way to go yet, I will not be feeling the satisfaction of work well done in my chosen profession, because there hasn't been any; but if we can make it out of the cold months in the comfort and warmth provided by my labour, I will consider it a success of sorts. I am lucky in this regard, as we are surrounded on three sides by well wooded farmland and, after windy weather, there is always dead wood to forage – often quite substantial pieces that can provide enough to burn for a week or two. The land belongs to the Farmer Friends and they regularly tell me where to look. We reciprocate their friendship when we can.

Recently they returned late one evening after an exhausting flight from Canada. We had left them Irish stew (consisting of parts of Milo and Donald) for their dinner, along with homemade marmalade and soda bread for the following morning's breakfast. We don't make a habit of welcoming them back from their holidays with food, but this was different. They had booked their trip in autumn but had hardly felt like going because a few days before their departure Female Farmer Friend buried her mother. She had buried her father only six months before – he had died suddenly aged ninety-four – and was not prepared to lose her mother so soon after. We both of course called in to the house to express our sympathy when we heard. The Beloved went to the removal and delivered a couple of trays of sandwiches for the inevitable stream of mourners. She was working in the city on the day of the funeral so I went alone.

It was held in Female Farmer Friend's childhood parish – an isolated place surrounded by long, low hills covered with pine and larch. The muddy field adjoining the church was packed with cars and a few young lads were directing people into available spaces. I parked and followed the gathering crowd through the small graveyard bordered by ancient cypress trees and younger, fuller leylandii. The large verdigris-covered bell tolled steadily as we filed into the old weather-beaten building with its squat grey walls and blue-slated roof. The rector conducting the service was a woman, as was the bishop who was in attendance, which surprised me but not, it seemed, the congregation. Irish country people are reputedly conservative, but they seem to accept progressive change without batting an eyelid. After the final hymn we followed the coffin the few short yards to the newly dug grave. She was to be buried next to her husband. The earth on his grave had not yet settled. I noticed the soil was a reddish brown and much loamier looking than the thick black clay around our hill. Around us were old, worn headstones covered in yellow and white lichen. Interspersed with these I could see the bright gold lettering on the clean stone of much more recent memorials. It was a cold, moist day, the cypresses were dripping with recent rain and the trunks of the few deciduous trees were covered in moss, like winter lagging. It took a while for everyone to commiserate and shake hands with the principal mourners, so I wandered, partly to keep warm but also to have a look about.

Some of the graves were decrepit: I could see one or two collapsed vaults surrounded by rusty iron railings; many of the grey slabs were time worn and illegible – the oldest I could read was inscribed in 1791. Many headstones commemorated four or five generations of the same family, all interred together. Many specified the townland of the deceased, saying things like 'formerly of' or 'died at'. This conjured up people who had moved maybe a few miles during their lives. There were inscriptions to those who 'fell in Flanders' – people who had travelled more than a few miles, but never came back. I noticed a recently erected pink-marble headstone for a man who had died in his ninety-third year. Just in front of it, in a little wooden frame, was a faded photograph of a tractor pulling hay. Underneath was written 'Dad, I will always treasure the days we worked together – till we meet again' and then the son's name. I wondered if he is still working the family farm where he and his dad once sweated together, or maybe he had flown home from afar to visit his father's grave. Old graveyards are strange, desolate places, but then burial is a strange, desolate ritual. At heart it consists of lowering a wooden box of cold flesh into a deep, damp hole and then walking away. I rejoined the thinning crowd and found the Farmer Friends. I gave them both a hug and left. Then I picked up the boys from their grandmother's, brought them home and cooked sausages for their tea.

We are in the depths of *gamh*, the dead half of the year, but there are signs of new life stirring. Clustered around the base of the big cherry tree, the pure white blossoms of the snowdrops droop from their green stems like plump pearls. The pointed young shoots of the daffodils are starting to poke up from the wet grass. And up at the farm the ewes have been brought in from the fields and shorn in preparation for lambing. The first lambs won't be born until March but shearing now reduces the ewes' bulk, which allows them more space in the sheep shed, and besides, being indoors, they no longer need the insulation of their fleeces. By the time they are back outside, the weather should be warmer and some of their wool will have grown back. The Farmer Friends don't do the actual shearing themselves – they hire specialists for the day – but they do shear their own cattle. This is not as absurd as it sounds.

When I made the game pie, I felt that I ought to leave a generous slice up at the farm. If I am honest, Male Farmer Friend probably shot most of its contents and it was only fair that he got to taste some. There was no one in the house when I called. I left the piece of pie on the kitchen table and crossed the yard to see if anyone was about. Something was happening at the cattle crush so I went to investigate. Male Farmer Friend was directing the beasts through the crush one at a time and plying electric clippers. He appeared to be shaving heifers. The ones that had emerged had a broad bald stripe in their hides. Running the length of the spine from head to tail, it looked like a reverse Mohican. He spotted me looking a bit puzzled and came over to explain. They were all in calf and, like the ewes, were being brought into the barn before giving birth. The bald patches were to keep them cool – a herd of cattle in a barn generate an awful lot of heat. They also generate an awful lot of crap, and as well as giving them tonsured haircuts, he was trimming their hairy tails to prevent them from getting caked in the stuff.

This year the sheep shearers were booked on the day Female Farmer Friend's mother died, and they just got on with it, unaided and unobserved, but I have seen them at their work before. They are a pair of brothers similarly built: stocky, broad shouldered and light on their feet. They arrive towing a trailer of gear and back it into the sheep shed. The trailer unpacks to form a platform with a gangway, wooden gates and swivelling arms from which the electric clippers hang. The brothers are cheerful and chatty but they don't hang around. Once they have set up, they set to. On a good day they can shear upwards of two hundred sheep each and it is quite hypnotic to watch them settle into a smooth rhythm. I was helping to clear the shorn fleeces out of their way and also to guide the freshly shaved and surprisingly svelte ewes into a separate pen, but whenever possible I hovered near the platform to watch closely as the shearers stooped and stretched. The brother nearest me was totally bald. I wondered if he sheared himself. Small beads of perspiration formed on his smooth scalp as reached for the next ewe. He took a firm hold of her fleece at the scruff of the neck, flipped her onto her haunches and pinned her between his legs. Then he ran the clippers in swift parallel passes down the sheep's side, starting along the belly and moving gracefully over the back. There were two small flourishes around the tail and the shoulders before he pivoted the placid animal on her rump and did the other side. The slightly yellowish fleece peeled away in one piece to reveal the white body beneath.

I was admiring this impressive combination of strength and dexterity when he spotted me and asked if I wished to give it a go. I knew I hadn't a hope of doing what I had just witnessed but I wanted to try. My ewe instantly sensed the hand of an amateur and began to struggle. It took most of my strength to hold her steady between my legs and I was sweating before I trimmed even an inch, but I managed the job. Bald brother was standing over me as he talked me through from start to finish. The fleece wasn't exactly in one piece as it came away and the sheep looked a little shabby. There was a sort of patchwork effect along her back and sides and there were a few tufty bits around the tops of her legs. I was reminded of a bizarre remark a friend once made when I bumped into him on the streets of London and observed that he had had a haircut. He said, 'Yeah, but it isn't finished!' My ewe didn't look entirely finished either. She wasn't nearly as neat and trimmed as her fellows. I wondered if she would be ostracised by the rest of the flock. As she entered the pen and joined the others, there was a general bleating. It could have been a greeting or it could have been laughter. Mind you, even the perfectly shorn sheep looked incongruous. Relieved of their wool, they look oddly naked and a little startled. Their hooves and heads look suddenly huge and, with their swollen bellies, the effect is slightly grotesque. Alongside these, my ewe appeared only a touch deformed. So all in all, I was pleased enough with my first attempt and relieved that I hadn't drawn blood. Bald brother plucked the next ewe in the line and got back to work. As he did so he gave me a smile and a swift nod – both hard to read. I was unsure whether the smile was encouraging or derisory, and the nod approving or dismissive. Possibly the latter in both cases, as this was a few years ago and I haven't been asked to shear a sheep since. I hopped down from the platform and continued clearing away the wool.

It is surprising how heavy and warm a newly shorn fleece is. We were packing them into massive sacks which were suspended from the raised forks of Farmer Friends' Loadall. This is a very versatile vehicle with interchangeable attachments for its hydraulic hoisting arm. The prongs of the bale-lifting-thingy were perfect for the job. The sacks were hanging from just above head height, and periodically someone had to climb up, drop inside and jump up and down for a bit to pack the wool tight. It looked like great fun and I wanted a go so I volunteered and my offer was quickly accepted. After I had been at it for a minute or so, I realised why others weren't keen to do it. It was hot and heavy work. The wool was warm and a little greasy. It was also less bouncy than I thought it would be, and after each jump down I had to heave myself up from the steel prongs of the Loadall. It was like doing pull-ups in the gym whilst heavily dressed, and after ten minutes I was dripping with sweat, itching and weak as a kitten. I discovered that I didn't have the necessary upper-body strength remaining to haul myself out, and I realised I would be trapped until I had packed down enough wool to raise myself up and step over the top of the bag. I was doubly grateful that the brothers were swift – the fleeces were coming thick and fast. When I finally emerged, I noticed that my previously grubby brown leather boots were sparkling. The oily lanolin in the fleeces, which normally acts as a water-repellent, had given them the best clean and polish they had had in years.

I took a breather and started gathering the fleeces again. The brothers didn't take a break till lunchtime. They epitomised speed, concentration and industry and the rest of us took our lead from them. As a time and motion study, the shearing was an object lesson in efficiency. The brothers sheared until there was nothing left to shear and our job was to service them. Ewes were led up the ramp and swiftly led away when done. Fleeces were cleared the moment they were dropped. It was a well-oiled, busy machine and everyone pretty much worked in silence. There was noise enough, though: the incessant buzz of the clippers; the constant thrum of the generator which powered them; and the steady bleating of the sheep. And it was hot. The weather outside was wintry. There was sleet, ice and bitter wind out there but we were oblivious as we worked intently in the warm fug of half a thousand sheep. At noon we trooped across the cold, wet yard and into the farmhouse for a hearty meal of hot boiled ham and floury spuds. Having your dinner in the middle of the day is a necessity when doing such work – you really need the fuel. I don't know how long we stopped for because no one was watching the clock. We ate, digested and chatted for a moment over a cup of tea then got back to work.

When the last sheep was sheared, there was a general sense of achievement and we cheerfully helped the brothers fold the platform back onto the trailer. Then they headed off over the hill. As I watched them disappear into the chilly drizzle I wondered where their next gig would be. It had been a one-night stand – to use a metaphor from my professional if not my love life – and I was reminded of what I have been told of the old 'fit up' days of the theatre.

Older actors I have worked with over the years have told me of touring with companies like that run by Anew McMaster, or 'Mac' as he was known. They would cover the country, occasionally playing in theatres, but more often than not they performed in barns, village halls or dance halls – possibly even sheep sheds: wherever they could set up a stage and gather an audience. A few rostra would form a platform, a curtain would be rigged and the company would evoke 'the vasty fields of France' or wherever. Then, like the brothers, they packed away the platform they had set up and headed off to the next town and the next show. I pondered the similarities between the lives of journeymen sheep shearers and itinerant actors and tried to avoid puns about 'being fleeced' when I noticed that my hands were velvety soft, as though I had been moisturising for hours. It must have been the lanolin in the wool. I also noticed that I stank of sheep. When I got home I was able to wash myself thoroughly, but not my leather jerkin, which I had been wearing. A permanent whiff of Eau de Mouton now forms part of its complex bouquet.

Yesterday I was driving home by the back road on the other side of the hill when I spotted a sleek grey sparrow-hawk swooping above a hedgerow. A few moments later I saw two big reddish-brown birds of prey flapping casually into the shelter of a large beech tree – the valley's resident buzzards. As I crested the top of the hill and passed Farmer Friends' big pasture, I noticed the earliest pair of this year's calves in the field with their mothers. And this morning I walked up to the top of our field, ignoring the sodden overgrown mess that is currently the kitchen garden, and saw the first yellow crocuses around the silver birch. Spring is coming.

IMBOLC

Rising Sap

Falling Snowflakes

'Now it's St Brigid's day and the first snowdrop in County Wicklow' as Heaney has it in his poem 'A Brigid's Girdle'. On this acre of Wicklow the white pearls of our snowdrops have opened. Their petals have unfurled and they look like clusters of tiny angle-poise lamps. The stiff green blades of the daffodils are lengthening and around the base of the silver birch tree, cream and yellow crocuses are peeping through. Winter is loosening its grip and the weather is improving slightly. Despite the still regular, drenching rain, we have had hints of a glimmer of sun – if not quite as much as the saint herself promised in the old saying: 'Every second day fine from my day onward and half my own day.'

For the last century or so, Patrick has been the pre-eminent Irish saint, but for hundreds of prior years, Brigid enjoyed a similar status. She was supposedly converted by Patrick, but swiftly became his equal. She was an extraordinarily influential woman who founded the Abbey of Kildare, and indeed she and her successors were seen as superiors general of all Irish monasteries. She was even consecrated as a bishop and subsequent abbesses of Kildare were accorded episcopal honours. And, of course, there are a thousand stories about her. One of the most enduring combines a typical mixture of magic and trickery. Wanting land to establish her convent she asked the king of Leinster for 'as much land as my cloak will cover'. No doubt amused, he agreed – whereupon four of Brigid's friends took a corner of her cape each and walked north, south, east and west. The cloak stretched and grew until it covered several acres. The king was astonished and less amused – however, he was impressed and he kept his word.

He gave her the land, along with money, food and supplies, and converted himself and his people on the spot. Brigid's, it seems, was a very practical magic.

About two hundred years after her death, an eighth-century monk named Cogitosus wrote about her life. He tells a story about a young woman who had taken a vow of chastity but had slipped down the path of pleasure and approached the saint with a swollen womb. The monk describes how 'Brigid exercising the most potent strength of her ineffable faith blessed her, causing the child to disappear without coming to birth and without pain.' The penitent girl then returned to health and virtue. I wonder how many of Brigid's successors as mother superior have yearned for the power to deal with naughty nuns in a similar fashion. This image of the saint as a sort of psychic abortionist lends a delicious irony to the much later custom of venerating chastity on Brigid's day. A girl nominated as 'the young Brigid' would parade around the town carrying a cross and wearing a white veil. Officially, the most exemplary virgin was chosen (although how this was established remains mysterious) but generally the prettiest girl was picked, and if her name happened to be Brigid it was an added bonus. The townspeople or villagers would follow the girl from door to door, praying and invoking the saint's benevolence for the coming year. She was a powerful protector and her orbit of influence was immense. She was patroness of cattle and dairy workers, chicken farmers, printing presses, midwives, poets and blacksmiths.

Her prestige began to wane in the twelfth century when her abbey at Kildare was sacked during a jealous quarrel amongst squabbling royalty. The King of Leinster abducted the wife of the King of Breffny, and he appealed to the High King of Ireland. The

High King deposed the Leinster King, who then requested the help of the English King, who joined in eagerly. In the ensuing ruckus, the convent was destroyed and its 'sacred flame' was extinguished. This 'holy fire' had been tended by nineteen nuns and kept perpetually alight since Brigid's death in 524. When the Abbey was reinstated after the royal scrap, the Archbishop of Dublin prevented the flame from being relit, ostensibly because the practice smacked of pagan superstition. In this he was right.

Saint Brigid of Kildare was an actual woman but her feast day and many of the attributes ascribed to her by Christian monks were those of a much older female figure – the ancient Celtic deity Brigit. Known as the Exalted One, she was guardian of cattle, medicine, arts, craft and poetry, and she represented fire, light and sunrise. The first of February was her festival and it was named Imbolc. The word derives from the old Irish meaning 'in the belly'. It evokes the swollen bellies of the pregnant ewes that still adorn the fields hereabouts at this time of the year. In terms of the solar year, Imbolc is a cross quarter day. It is the midpoint between the winter solstice and the spring equinox. In old mythology Brigit was the daughter of Dagda, the 'Good God', and she married Lugh, the sun god. She was a powerful fertility figure who presided over the living half of the year and Imbolc represents the beginning of her ascendancy. She was also known as Danu or Dana, and according to myth, the goddess Danu, or Brigit, conquered a much earlier female deity: An Cailleach, the Divine Hag of Winter.

The Divine Hag was known as the Veiled One. She was an earthier, more ancient sacred figure and she was imagined as a spectral old woman. These were days when life was harsh and probably short. Age commanded respect – it was venerated, and an elderly crone, far from being a foolish, weak creature, represented strength, wisdom and power. The Divine Hag built cairns, mounds and megaliths and she formed the landscape. She froze the ground with a blow of her staff and she beckoned the wind and the rain. Storms in February were caused by flights of her people riding wolves and wild pigs. She owned a great bull whose bellow impregnated every cow that heard it, which was perhaps not in the bull's best romantic interests. In some traditions the Veiled One was mother of all gods, and she personified the glory and the destruction of the natural world. Nothing escaped her control … until the Exalted One came, bringing the light and the warmth of the sun.

Unfortunately Brigit can often be a little tardy – her day this year was cold, wet and gloomy, although this, apparently, is a good sign. If she wished, the Divine Hag could delay the onset of spring. When she chose to do so, she arranged good weather on Imbolc so that she could collect enough firewood to get through the cold times ahead. If the day was fine on the first of February, it meant that she was out picking sticks and had therefore ordained a long winter. If the day was foul, then she was asleep and warmth was on the way.

This ingenious story encapsulates a typically Irish ability to remain paradoxically optimistic about the weather despite continuing evidence to the contrary. The weather is terrible – which means it's going to be good! The Beloved's mother gave us a St Brigid's Cross when we bought our house. Woven from straw, it hangs on the kitchen door as a token of good luck. The goddess Brigit floats behind the saint and, beyond them both, the Divine Hag hovers. Since coming to live here, I have become more mindful of the beginning of spring, Brigid's day, Imbolc.

But these layers of significance have all been superseded. For the past nine years, the first of February marks the anniversary of day I became a father.

The boy was born at ten past six in the evening after the Beloved had laboured for thirty-nine and a half hours. For most people this is a working week. She didn't know what hit her and I am still in awe of what she did. Having a baby is, at the same time, the most ordinary and the most extraordinary thing. When I cast my mind back to the birth of either of my sons, I remember mundane details with electric clarity.

Older Boy's arrival began in a gentle enough fashion. I awoke in the small hours to find the Beloved padding about in slippers and dressing gown. I got up and made her a cup of tea – she was pretty sure it had begun. We were in the flat we used to rent in town. I started to get things ready and she spoke to the midwife. A few hours passed calmly but quite quickly. I began to fill the birthing pool we had hired and which I had erected the day before. She wanted a soothing, peaceful environment, so I lit candles around the room and played gentle music. In due course the contractions increased in frequency and intensity, and the midwife arrived. I slipped into my role, which was to be present without being annoying. I wasn't entirely successful in this. At one point, during an especially intense contraction, the Beloved caught sight of me looking at her with my best concerned, empathic expression. This was apparently infuriating. So I tried to keep my face blank and open to whatever she wished to see in it. I know actors who have made a fortune doing this. It actually requires a high degree of skill, which I don't possess. Luckily the Beloved was far too occupied with her own mammoth task to be paying much attention to me, so she didn't for the most part notice my barely suppressed astonishment at what she was doing. Especially the noises she made. I have sat at the back of the London Coliseum and heard a soprano fill that huge theatre with a pure, piercing cry. I have heard the competitive clamour of neighbouring bulls as they bellow all night in fields adjoining ours. And one night in the African bush I heard the resounding roar of a distant lion echo around the hills of Kwazulu. These are like so many sparrow farts compared to the sounds she made that night. They came from the centre of the earth and were glorious, if a little terrifying. I nearly laughed out loud, not because I was in any way amused, but because I was stupefied by what I was hearing. I stifled any inappropriate noises by biting my hand and looking across the birthing pool at the midwife, who was smiling and nodding approvingly. She was younger than both of us but wise beyond her years. A tall woman with a shock of curly, reddish-brown hair, and exuding warmth, competence and reassurance, she was a modern variation on the old Celtic earth deities. Aware of all the latest ante-natal research, she also had the practical experience. She said if things continued as they were going, the baby would come before midnight.

At one point the Beloved said she was getting a whiff of tobacco smoke which was making her feel nauseous. I went outside to investigate. Behind the flat and under our open bathroom window, three girls from neighbouring flats had just lit up and were enjoying a sly fag. I felt bad for disturbing them but there was an implacable imperative to be obeyed. I asked them if they wouldn't mind smoking elsewhere as the Beloved was having her baby right now and the fumes were upsetting her. At that precise moment, she let out a long, loud yowl that dwarfed all the previous ones. The girls' mouths fell open and their cigarettes dropped from lifeless fingers. They scuttled away sharply. If that didn't impress upon them the need for contraception, nothing ever will.

Then things sort of stopped. Basically the boy got stuck and the Beloved laboured for some exhausting fruitless hours. She got into bed and withdrew deep into herself. I knelt beside her and whispered into her ear. I did my best to concentrate all my love and admiration into soothing, rhythmic phrases. She toiled silently far away. Then the midwife called it.

We went to the maternity hospital and the whole procedural machinery kicked in – an epidural, a foetal monitor and, twelve hours later, a delivery involving a trolley laden with steel tools and an instruction to 'Look away, Daddy!' When she and the tiny boy were taken to the neo-natal ward, I was allowed to linger for a while, but the hospital rules meant I eventually had to go. I went back to the empty flat. The water in the birthing pool was cold and needed emptying – that could wait. The pillows were piled high on the unmade bed where she had sweated and struggled. Unlike her, I was physically unscathed, but I hadn't slept for forty-five momentous hours. Tears trickled down my face, I'm not sure why. Relief maybe, exhaustion possibly, pride and awe without doubt. But also, perhaps, my first glimpse of a deep, abiding joy. She was pretty battered when I saw her the next morning. The epidural had long since worn off and she was wrung out, in pain and struggling to breastfeed the boy. He, of course, was amazing and beautiful – despite the facial scraping and bruising from the forceps. We were both eager to get home, although the hospital was less keen to let us go. She hadn't eaten for two and a half days, so we waited till lunchtime. A most unappetising plate of overcooked lamb and dried potatoes was offered. She snorted it in seconds and we left. She had had two labours: the one at home that she had wanted and which went on for long enough in itself; and then a second one at the hospital. She felt as if she had been run over by a truck and, as soon as she was able, I brought her home to Wicklow.

For the next few days she and the boy slept and I … well, I kept watch, I suppose. The clock was irrelevant. The quiet hours passed and time unfolded with an easy inevitability in the snug house. I remember a specific but representative moment. They were asleep upstairs under the snowy-white down duvet. In the sitting room the stove was lit and steadily glowing. The afternoon darkened and snow began to fall on our field. Through the window I watched the soft, feathery flakes descending. Once in a while they flurried and swirled, mimicking the patterns of the flickering firelight. Everything rhymed. I lay on the couch drowsing and utterly contented. I let myself fall asleep, acquiescing to unconsciousness in the certain assurance that all was well. When the Beloved was ready she decided to go outside. Brigit had not quite superseded the Divine Hag and, despite the yellow sun, the morning was cold and crisp, so we dressed ourselves warmly. She wrapped the boy and swaddled him to her body in a sling, then we left the house and walked up to the cross. Sliabh Buí and the surrounding hills were snow-capped. The fields around us were dotted with fat, pregnant ewes. Our boots crunched wafers of ice and our breath made shapes in the silence. Up at the cross we met the Farmer Friends, whom we were just getting to know. They asked, 'Has that baby not been born yet?' The Beloved opened the top buttons of her thick coat to reveal the small, pink face asleep on her breast. They admired him and congratulated us. Then we chatted about babies and the weather and the imminent lambing season. We walked back down the hill and I felt the earth revolve beneath my feet.

Now he is nine years old and lengthening. We measure both boys on each of their birthdays every year and make a mark against the wall beside the kitchen door. Our heights are also marked – in gold, like targets. I am just over a foot taller than he is now and his mother has eight inches on him. He will probably overtake her at eleven and I expect him to pass me in his very earliest teens. He can't wait. He thinks, and I remember believing the same, that teenage years bring glory. I hope he is not as disappointed as I was. His first birthday bewildered him. I have a photograph of the day – actually I have hundreds, but there is one in particular. He is sitting in his high chair wearing a party hat perched at a jaunty angle. Behind him are balloons and streamers. He is flanked by his grinning grannies and on the table in front of him is a cake with a candle. He looks utterly perplexed. He hadn't a notion of what was going on with the deranged adults around him.

He got the idea over the next two years but his fourth birthday was fraught. It was the first year he invited friends from playschool and his mounting excitement was palpable. We cocked up the starting time on the invitations – everything was ready at two, but the guests didn't arrive till two thirty. It was the longest half hour of his life and by the time the party began he was pumped up and ready to burst. Twenty minutes in he declared it 'the best party ever' and thirty minutes later he was overwrought and sobbing. The breakdown came as the Beloved and I were in the middle of a puppet show which we had hoped would wow the children. I was voicing a sly fox and a deliberately unfunny clown – she was an evil crow and a witty princess, which she was particularly relishing. A confrontation between the clown and the crow really upset the crowd and caused a riot. The birthday boy just wanted it all to stop. To settle things down we had to reluctantly abandon the show, both of us disappointed not to get to the terrific finale we had planned. Later, at the table, when the children were scarfing cocktail sausages and cake, the boy couldn't handle the gabble and clatter and he kept pleading with everybody to 'be quiet!' Naturally enough, he kept being ignored. The subsequent party games sustained the general level of giddy intensity, despite our best efforts at exerting a calming influence. It was fucking exhausting.

Strangely, he went to bed that night happy and even thanked us both for a great party. God, I love him. He had entirely forgotten his earlier anguish. I couldn't really say the same for myself, though. Later that evening I found myself slightly unbalanced by the emotional turmoil of the day. Sitting outside on the bench facing Sliabh Buí, a relaxing cigar was failing to smooth the turbulence I couldn't deny feeling. I was confused by my confusion. As is often the case, the Beloved helped me figure it out. And as is often the case, it was not about the boy, but about me. I have missed having my father around since I was nine. I suppose my stepfather did his best to display affection for the unhappy boy he acquired when he married my mother, but I don't think he really understood how to feel it until he and she had children of their own. We were never close and didn't get on easily together until I had grown up and left, and even then, it was a guarded relationship and there were pitfalls. He died when I was thirty. My dad and I have had challenges over the years but we have surmounted them; we don't see each other often yet we have achieved a mutual acceptance of the past; we enjoy an affectionate conviviality; and we love each other. But there are large gaps in our shared history and he is distant. We have lived in different hemispheres for more than four-fifths of my life.

When I became a father, I had no doubt that my mum would become an integral figure in my sons' lives even though she lived in another country. During the Beloved's long labour I phoned her to say things had begun. When I rang her later that day to say 'It's a boy!' she was at the airport, boarding a plane to Dublin. I suppose I hoped that my father and stepmother would become equally important to my boys, and somehow, in my submerged mind, I thought that a deep connection between grandfather and grandsons, through me, would fill the hole I felt for so many years of my life. Of course it can't and indeed it hasn't. Trying, however subliminally, to recoup ancient losses has only made me stumble in my steps into fatherhood. Another hazard has been fragmented memories of a childhood fear that, in some confused sense, it was my fault that my father left. On Older Boy's fourth birthday, I felt in some equally vague way that it was my fault that he got so upset. I have learnt to relax a little. I don't blame myself as much as I used to for a lot of the pain and trouble of my early years, and I have tried hard to rid myself of a subconscious belief that I ought to feel bad – that I don't deserve a happy family. Now my older boy has reached the age I was when my childhood home broke apart and everything changed. It was a milestone in my life. I look at him now and I see myself then. I frequently rediscover aspects of my younger life in his. Sometimes his burgeoning emotions fuel my remembered ones and once again I am ambushed by boyish fears and worries which rise up in me, renewed and vigorous. I am learning to recognise them for the insubstantial ghosts that they are. I am aided in this by the

Beloved's perceptive eye and by the security I feel in the certain knowledge that I am not going anywhere.

The boy's subsequent birthdays have been fun-filled celebrations, largely because we have learnt to calibrate his parties to meet his wishes, not our ideas. The following year we took him and his pals to a nearby bird-of-prey sanctuary followed by a picnic – it was a fine day: the Divine Hag must have been out picking sticks. We have also been to the pictures and the bowling alley, and this year it was a giant trampoline followed by table tennis, pool and a plateful of chicken nuggets and chips. My own birthday follows his by a fortnight and is therefore usually overshadowed and not as much of an event as it used to be. For the past few years the boys have been old enough to make personalised cards, which I love, and this year they also gave me underpants and socks. My other gifts were a new jumper, aftershave and a voucher for a store in town where I 'can get something nice for myself'. I need make no further comment. The Beloved had planned for the two of us to get out to the pub for an hour in the evening, but that fell through for reasons too dull to record. So my birthday this year was not especially notable, apart from the fact that it had snowed early in the morning, which made the day beautiful. When I opened our bedroom shutters the light that bounced in was unexpectedly dazzling. A bright sun shone from a blue sky and the field was stark white. The pristine beauty moved me with a sharp, surprising thrill and in the freshly fallen snow even the derelict hen house looked clean and new. The first snow of the year is always exhilarating. The boys feel the same delight, but much more intensely.

When the older one was six we were standing by the stove looking out the window at the fat flakes settling on the piggery.

He said, 'Dad, I'm so happy it's making my eyes water.' That particular winter was a frozen one. There were blizzards and ice and our road was practically impassable. By the time the snows had receded, the boys were heartily sick of the cold. They have forgotten this, of course: when they saw the white blanket draped around the house on the morning of my birthday, they were delirious with excitement and we were soon outside. Regrettably for them, the fall had not been heavy. There wasn't enough for a decent snowball let alone a snowman, but the acre looked unblemished and reinvigorated. With all the recent drizzle, rain and mud, the field has had a tainted look. Drab colours of damp grey and marshy brown have been predominant, but now the grass was poking up between the last patches of snow in bright-green tufts. This was no consolation for the boys – by the time they got back from school it had mostly melted. They were sadly disappointed. They can still vividly recall those heavy snows of two or three years ago and they yearn for their return. They remember whole families of giant snowmen and snowball fights that went on for days.

One morning during that icy winter Male Farmer Friend arranged to meet us in the high pasture up behind our house. We took our time trudging there: the snow was deep and the boys were smaller so progress was slow. The silence was thick and broken only by our panting breaths and the muffled squeak of our footsteps. Male Farmer Friend arrived on his quad bike and rode it in a diagonal line across and down the sloping field to make a long furrowed run, perfect for sledding. We had no sleds but he unveiled his plan. He had brought a few heavy plastic feed bags stuffed full of straw and with loops of stout twine tied tightly to the tops. If you sat on the bag with the top between your legs, then gripped the loops firmly and leaned back, all you needed was a little shove in order to fly down the run at Olympic speeds. The further back you leaned – the faster you went. The first few times the boys were clutched in our laps, but they soon got up the courage to fly solo. Once in a while someone lost control and ploughed into the soft snow on either side of the furrow, which was possibly even more fun. Eventually even the boys grew tired and they followed their mother back to the house for hot chocolate. I rode pillion on the quad bike and accompanied Male Farmer Friend up to the top of the hill to check on some sheep.

The view was staggering. The low winter sun was unobscured by cloud and the eggshell-blue sky seemed limitless. Just beneath us to our left, the deep, dark pine forest stood silently under its mantle of glistening icicles and silver-grey frost. Further down in the valley, we could see the white rooftops of the village clustered around the glinting spire of the church. The snow lay all around in every direction. The vast white expanse blurred the field boundaries. We gazed away across the blank white open reaches to Sliabh Buí on the far horizon. We didn't actually check on any sheep. We just stood in the high place, knee deep in the snow, breathing the clean air and looking at the quiet, shrouded landscape stretched below. It was beautiful, but after a week or two of similar weather it all became a pain in the hole. Our road was buried under a deep drift – we had to hitch rides from passing tractors or quad bikes to carry anything heavy from the village shop. The water pump in the garage froze, so we were unable to run a tap or flush the loo for quite a few days. We thawed the pump with a borrowed blowtorch and hung a lightbulb next to it. We kept this lit for the duration to keep the pump from freezing again. A single lightbulb doesn't throw much heat, but it was just enough to keep the temperature above zero.

This year there has been no sledding and, apart from the one visit to the ice rink on New Year's Eve, the only winter sports we have engaged in have been vicariously via the television. The younger one turned out to be especially keen on figure skating and may possibly have a talent for it. We have spent an entertaining hour or two watching him demonstrate how it should be done. He ice dances around the sitting room, leaping and twirling, and he seems to understand the difference between a triple Salchow and a triple Lutz. Combining this recently acquired expertise with one of his regular enthusiasms, he has invented a new sport. He gets his scooter up to top speed in circuits of the school playground and then throws in some moves straight out of ice dancing. He extends a leg to point front or point back whilst holding the opposite arm over his head or out to his side with balletic grace. Then he elegantly sweeps the extended arm back to grab the handlebars and extends the other whilst simultaneously switching legs. He finishes with a bow but keeps scooting. One fiercely complicated manoeuvre involves twisting his upper body to the right and his lower body to the left whilst at the same time wrapping one leg behind the other to point forward, without losing speed. My favourite is a very risky hands-free stunt. He leans forward, resting his chest on the handlebars, and holds his arms out and back like he was doing a swallow dive. At the same time he holds a leg straight out behind and glides serenely by. It is really quite impressive and he calls it Figure Scooting. He is not as tall as his older brother was at the same age and he is maybe a little shorter than the average six-year-old, but what he lacks in bodily stature he makes up for in general size and presence. He is a Ninja tornado so we were a little surprised to discover his love of dance. One afternoon last year when home sick from school, he was paying rapt attention to something noisy on the TV. The Beloved shouted to him from the kitchen, asking what he was watching. 'Nothing,' he replied. 'Don't come in!' Naturally she couldn't resist peeking. He was totally transfixed by a Bollywood musical but didn't want to be caught watching it. He found it simultaneously alluring and embarrassing. He is perhaps a bit young to experience that mixture of guilt and desire with which I am so familiar due to my relationship with cherry liqueur chocolates.

Dying Calves

For weeks now, the rain has been almost incessant. The lower end of our field was a marsh, but now it is more like a swamp, and there is a gentle but steady stream of water bubbling out through the bottom gate and into the ditch at the side of the road. I have avoided walking down there despite the pleasing sound of the trickling flow. Up at the top end of the field, the kitchen garden is looking grim and I still haven't planted any garlic. We have dug the last of the potatoes, which we have been eating since July, and now we must start buying them again. All of the crocuses are now out and under the silver birch there is a swatch of purple amongst the white and yellow, but the pretty effect is almost obscured by all the dead branches that have come down. The wind has been fierce. The Divine Hag may not have smitten the ground with her staff to bring ice and snow, but she has been riding on wolf-back with her people because we have had the worst storms for years. Hundreds of thousands of trees have fallen all over the country and power lines are down everywhere.

Outlying parts of the village were without electricity for days. Most mobile-phone networks, as well as land lines, were down. This meant that you had to actually call in personally to see how someone was doing. I brought a flask of hot coffee over to a friend who lives alone up a quiet lane. Access was blocked by a fallen beech tree, but other neighbours were already attacking it with chainsaws to clear her way. In a bit of a crisis, everybody seems to go out and about to check on each other. We got off lightly and were only without electricity for a day and a night so the boys didn't have time to grow tired of the novelty. We had candles, wood and a gas hob, so we were well fed, sufficiently warm and flatteringly lit. There was no television or Internet access, nor were we sure when we might be able to recharge our phones, and the boys had to accept that there was to be no screen action of any kind. They were only briefly annoyed and swiftly accepted the fact that we had to entertain ourselves. Younger Boy decided to continue his exploration of the world of dance. He instructed me to hum a ballet for him to throw shapes to. I had a stab at *Swan Lake* and then quickly segued into the *Nutcracker* suite. Thereafter I threw in pretty much any classical tune that occurred to me – what did he know? He flibbertigibbeted around the place until his brother joined in. The candles flickered to their cavorting – great leaps, arabesques and jigs – as they triumphantly matched anything I could offer. The high point of their choreography and the crescendo of my lilting was a jubilant rendering of 'The Blue Danube'. They danced by the fire and their projected shadows vaulted across the ceiling. They were quite aware of the potent impact of this lighting effect and they relished it. Their capering was exaggerated into a reeling chiaroscuro – it heightened their pleasure and inspired greater efforts. I hummed myself hoarse trying to rise to their challenge.

Bedtime stories that night felt more intimate than usual. The limited spill of the soft candlelight concentrated the eye on the page and the doings of Hiccup the reluctant Viking and Horrid Henry seemed more vivid. We have become so used to the flattening effect of electric lighting that our senses are dulled. Our imagination is no longer sharpened by the contrast between darkness and light, and we have forgotten the dramatic possibilities of the ordinary. The following evening when the electricity supply had been restored, I suggested that nonetheless we should give the ballet by candlelight another go, but the boys opted for their gaming console. The Farmer Friends had no need of candles. The Beloved had called in and found them snuggled up watching the telly, which they were running off the battery from a tractor. Mind you, this was not the screen they should have been ideally watching. The power outage had disabled the camera in the cattle shed. They have had this rigged in order to keep a benign but unobtrusive eye on the calving. In previous years I have seen newly born calves as young as an hour old. This year I helped deliver one.

It was a rare fine day and the ground was soft – in other words a perfect opportunity to start digging the vegetable garden in readiness for sowing and maybe even plant some garlic. So instead, I went up to the farm to see if anything was happening. Female Farmer Friend called me into the cattle shed to see something interesting. In one of the side pens was a cow pacing restlessly through the fresh straw with a pair of small hooves sticking out her back end. The calf was about to be born and did I want to stay and watch? We stood at a decent distance to observe.

Cattle like to be left alone and mostly manage by themselves – intervention can arrest the birthing process and should only be attempted if there is a danger to the calf or the mother. They are big and occasionally violent animals, as I remembered, so we kept well back. Usually one front hoof emerges first, then the other and then the head follows shortly afterwards, tucked down between the extended legs as if the calf was diving out. Once the legs, head and shoulders are out, gravity helps the rest follow. Sometimes one of the hooves is not extended fully and its tucked-back tip can impede things, in which case you flip the bent hoof till it is pointing straight and out the calf comes. Male Farmer Friend joined us. The protruding hooves had withdrawn a little and he decided to help things along. He grabbed a loop of rope and beckoned me to follow. We slowly approached the cow, who circled to keep her head towards us and her rear end protected. This minuet continued for some minutes, but we moved slowly and calmly until the cow relaxed and we got

around to the business end. Male Farmer Friend looped the rope around the now nearly disappeared hooves and told me to grab hold and pull. I started to tug and yank eagerly, which was not ideal. The cow didn't complain or question my technique, but Male Farmer Friend did. He told me to apply a firm but consistent pressure. I leaned back and hauled steadily, like in a tug o' war, and the calf was born. Out came the legs, then the head – dimly visible through a milky grey membrane. Then the slick shoulders, the body and the back legs slithered out and the calf flopped to the ground. The mother turned and, ignoring us, began to lick her newborn thoroughly. She also ate what remained of the membrane. This is apparently an instinctual thing: her undomesticated forbears did this to dispose of the evidence of a birth, thereby keeping potential predators unaware of the existence of vulnerable prey. The licking also stimulates the new calf and encourages it to move. The calf had been stuck for a while and the Farmer Friends were a little anxious about its survival chances. It lay still in the straw for a moment or two. Male Farmer Friend poked some straw into its nostrils to irritate it into snorting and thus get it breathing properly. With the body licking and the nostril poking, the calf stirred and began to look less like a deformed otter and more like its mother. The next day Female Farmer Friend sent me a picture of 'my' calf – or Philomena, as she had named her – placidly following the other new arrivals out of the shed and into the fields.

Not every assisted birth is as simple. Sometimes a mechanical aid is required. Male Farmer Friend showed me the metal contraption he uses on such occasions. It has a frame like the outline of a large bicycle saddle, which fits over the rump of the cow. Attached to this is a hook and something resembling a car jack. The calf is winched out. It sounds, and indeed looks, brutal but it often saves the lives of the calf and the cow. It was used a few days after Philomena's arrival to deliver a pedigree bull calf. A fine big thing, he survived his traumatic breech birth at the practised hands of the skilled vet, only to die inexplicably a few days later. The morning after that loss I was driving with Male Farmer Friend across back roads to collect his car from the mechanic he uses. I was to drive his pickup truck home after dropping him, and the dead bull calf was in the back so he told me the whole story. He looked very tired and a little low. He admitted to not being in the best form but said there was something I should see. We were taking the carcass to – there is no more appropriate term – the knacker's yard.

When we arrived, we backed up to the high open doors of the main concrete shed. I got out and noticed, in an outhouse, a chest-high pile of something nasty looking. These were whole cow hides heaped on top of each other and covered in salt. Viscous pink gloop was seeping down the side of the pile and across the concrete floor. There was a cloying, unpleasant smell. I tried not to breathe through my nose. I went to the back of the pickup and we uncovered the dead calf. The head looked pretty. It was unblemished, chestnut coloured, and the eyes were closed as in sleep. From the neck down it had been skinned and the pale naked flesh was almost translucent. The pelts have some value so Male Farmer Friend had flayed it. A small, cheerful man in blood-spattered overalls appeared. He chirruped a greeting, pulled the calf from the pickup and flung it on a nearby pile. Some of the intestines spilled out. Various things shouted for my attention. The pile consisted of other animal bodies – more calves, some sheep and a couple of dead dogs. To my left was a rack hung with hooks like those I had seen in the abattoir. These were for hanging cattle from to facilitate stripping their hides easily. Beyond that was a large skip full of skinned carcasses. Another small man in blood-spattered overalls was filling this with more corpses. He was using a JCB earth mover and he waved and smiled at me as he drove by. Directly opposite me was a huge, orderly pile of dead cows. They were stacked regularly in layers along the back wall. They had been decapitated. The heads were in a neat line along the floor in front of the bodies. It was an oddly formal detail in the midst of the shambles.

It took a while for all this to sink in. Male Farmer Friend had wandered off and left me. He needed to deal with the paperwork – every animal, wherever it ends up, has to be accounted for. The first cheerful small man was now sweeping blood into some gullies cut into the floor. I asked him about the neat row of chopped heads. He chortled and explained in keen detail. It was to allow the vet access to the brains in order to test for mad cow disease. If evidence was found, it could be traced back to the herd of origin and, hopefully, controlled. Thankfully, the disease is very rare. I asked him what happened to all the carcasses after being tested and skinned. 'Ground into bone meal to fuel German power stations,' he told me with a proud grin. His teeth were dazzling white, which seemed incongruous. I found myself staring at them in fascination – perhaps I didn't want to look elsewhere. I asked him a final question: what were the most common causes of death amongst the animals in mounds around us? 'Oh, anything at all!' he said brightly. 'Just like us!' I said I hoped we wouldn't end up being thrown onto a charnel heap. He said, 'What would it matter to you if you're dead?' Then he winked and walked away whistling. I know people who work at their ease in comfortable, clean, sweet-smelling surroundings and are miserable. The two cheerful, small men in the knacker's yard were the happiest I've met in a long while.

Still, I hope Philomena will not end up passing through their hands. She has not been well. Meningitis was suspected and the vet was called. He noticed that she was grinding her teeth and losing her sight – classic indications of lead poisoning. Male Farmer Friend was thoroughly perplexed – there was no lead on any of his land – so they investigated. In a corner of the field where the new calves were grazing along with their mothers, they found, near the roadside hedge, an old battery about the size of an eggcup. Its casing had worn away, exposing the lead, which had subsequently oxidised to form a salty-tasting crust. Cattle love a salt lick and the best guess is that the calf must have found the battery and given it a good suck before spitting it out again. The vet accordingly gave her an antidote for the poison, so hopefully she should survive. I assisted at her birth, so I don't really want to hear of her being flung on a heap prior to be being ground into bone meal. I would much rather imagine her pan fried, grilled or slow roasted with carrots and onions.

Returning from the bull calf's grisly end, we took back roads unknown to me. In nearly every field we passed there were clumps of men with chainsaws surrounding large fallen trees brought down by the February storms. This morning I pondered this sadly sobering fact whilst sitting in the sunshine sharpening my axe. The daffodils are starting to come up, the buds on the big cherry tree are swelling and some nearby blackthorn trees, always the first to blossom, are beginning to flower. The wettest February since the last recorded really wet February has turned into a fine and dry early March. I really, really ought to start preparing the vegetable garden, but despite the unfamiliar warmth we still need to keep the stove burning so wood must be collected. The trees are down – they may as well be used.

We have only felled one tree on our acre since we came here. It was an old leylandii growing in the bottom corner of the field away from the road. There were two of them planted tightly together. They made a single, giant evergreen bush which made life very difficult for a stunted whitethorn struggling to survive beneath them. Taking one of them out opened the view to the eastern hills, freed up the thorn tree and gave us firewood for two years. They are the least attractive of the evergreens but I still felt a quantum of guilt for cutting down an otherwise healthy tree. However, as an act of contrition we have since planted an orchard, a birch and two ornamental cherries, amounting to ten trees in all, so I think we have balanced the books. I do my best to be ecologically aware and generally regard green projects favourably, which is why I try to see the wind turbines on the flanks of Sliabh Buí as a good thing, but they are proliferating. A few years ago, there were two or three turning prettily on the distant horizon but now there are more. As I sat under the cherry tree applying the finishing sweeps of the whetstone to the axe blade, I counted them – there are seventeen dotted along the skyline. They are not yet an eyesore, but I don't find them as aesthetically pleasing as I once did. Now they imply a threat of something ugly. I suspect notions of beauty are often subconsciously informed by one's ideological attitudes. I remember talking about windmills to someone at a party. He was making 'jolly' remarks over dinner about Arabs and lesbians – we didn't see eye to eye – and he thought windmills were a blot on the landscape. I claimed they were peaceful and pleasing to look at. It was politer than calling him a bigot. I imagine he saw them as places where *Guardian*-reading, vegetarian earth mothers danced around in Sapphic rituals and therefore to him they were hideous. I saw them as an energy source free from the clutches of oil-rich, warmongering Texans and therefore beautiful. They don't look so pretty to me now. They are still making green energy, but this electricity needs to be conveyed away and the cheapest way to do this is via huge pylons. A semi-state body is planning to build a corridor of these things along one of four suggested routes. Our house is directly in the middle of one of them. The decision has yet to be made or has, I fear, been made but has yet to be announced. Full of scorpions is my mind when I think about it. We have protested, of course, and there are anti-pylon posters all over the village and beyond. We have written to the body concerned and been ignored. Of course there are large issues and big pictures and long-term energy requirements et cetera, et cetera. But simply and selfishly put, it is quite possible that the beautiful corner of Ireland I inhabit could be blighted.

I also rant and rail about the semi-state nature of the organisation concerned. Will the 'state' part enable the project because of perceived national needs and will the 'semi' part absorb the profits? Impotent fury is good for neither body nor mind and makes me disagreeable to live with, so it is best not to think too precisely on the event. I try to ignore what feels like an axe hanging over our heads – far better to swing my own freshly sharpened one and imagine I am splitting something other than a log. Far better, also, to concentrate energy on achievable goals rather than rage emptily at irresistible forces like a whelk whining at a tidal wave. Or so the Beloved says. Annoyingly, she is right. I cannot stem the flow of global corporate interests but I can take salsa lessons.

All four of us are currently engaged in various village activities. The boys are learning Taekwondo; the Beloved has joined a choir; and I have been entered in a disco competition. This is entirely the fault of the GAA. The village's senior club is in dire financial straits and the spearhead of their fundraising strategy is to be a dance tournament. As I am vice-chairman of the juvenile club, and also because I am an actor and therefore (it is generally believed) not averse to making an arse of myself in public, I have been roped in. The big night will be in early April, which is getting uncomfortably close. We have had a few weeks of lessons now and the competition is getting a little more intense. People are beginning to take it seriously – particularly my partner. She is the daughter of my chairman and she feels the honour of the juvenile committee is at stake. She is also a good mover and, given that she thinks she can whip me into shape, she is confident that we could win. I enjoy a dance and like to think that I can body sculpt with the best. But I make it up as I go along. Consequently I am finding it difficult to learn the precise steps, and even when I do grasp them for a moment, repeating them seems to be impossible. However, my partner is patient and I am starting to enjoy it. But time is running out.

The club has engaged a professional instructor to guide us, but all the dancers are amateur. Most of those who have agreed to do it have done so because they thought it might be fun. People usually only enjoy an activity if they can actually do it, so most of the competitors are competent movers, which has encouraged our teacher to throw us in at the deep end. We were straight into salsa, mambos and the cha-cha-cha. A pattern quickly emerged for me: I would spend the first half of the session deeply frustrated with myself for dancing like a pissed hippo; then I would begin to enjoy myself and start to get it; and then it would all fall apart again at the beginning of the next lesson – a typical journey, no doubt. I am beginning to get the overall shape of our routine, but remembering the detail is still tricky. A week ago I had what I thought was a bright idea – using my phone to record my partner and me doing the steps correctly, which I could then use as an aide-memoire to practise at home. This was not wise. Everyone has a self-image based on the evidence of their own eyes: we see ourselves in mirrors on a daily basis, whether it is to shave or to apply make-up. These separate activities are not necessarily limited by gender – look at drag artistes, or East German shot putters. We think we are seeing reality. But most of the time we instinctively look in a mirror in such a way as to present our 'good' side. We tilt our heads this way and that until we accentuate our best features. We flatter ourselves. As an actor, I am often on camera, which supposedly never lies, but it can fib a little – I am always assisted by make-up, lighting, camera angles and so forth to make the best of a bad job. And on set I am frequently, untruthfully, told that I am 'looking good!' So I had an interior picture of what I looked like dancing. Also, I was growing in confidence as I was becoming more familiar with the steps. I imagined my phone had recorded an elegant older man twirling a young woman with a deft grace and a crisp style. The reality was devastating. The camera can occasionally lie. A mobile phone never does.

I didn't look at the clip till the morning after that lesson. The weather was glorious for early March. The sun was shining and the daffodils were ablaze. Shrove Tuesday had recently reminded the boys of their taste for pancakes, so I fried and expertly tossed some for breakfast. As we walked down the hill to school, I thought I would have a quick glance to refresh my memory and keep the routine in my mind's eye. I took out my phone, found the file and pressed play. I saw a gauche, fat, balding, middle-aged knob lurching around with a girl nearly young enough to be his grand-daughter. Apart from the horror of my clumsy twisting and the ghastly wobbling of my paunch and jowls, what was irrefutable was my age. I often joke to friends that the inner 'real' me is thirty-two, and that the extra couple of decades are merely accessories that I can choose when or when not to carry. Looking at this old git, the decades, and more, became heavy iron chains. I slouched down the road behind the cheerful boys. The bright sun hurt my eyes. The twittering birds were annoyingly noisy. The fluttering daffodils looked kitsch, gaudy and ridiculous.

The club has issued us all with questionnaires to fill in. They want witty answers to predictable questions which will make up an entertaining programme to be sold on the night. We have been asked to supply amusing nicknames. My partner has settled on 'Lollipop'; I have yet to decide. I may leave it a week or so to allow my self-image to recover, otherwise we may be billed as Lollipop and Lardarse. I certainly won't be calling myself Twinkletoes, as one of the tittering girls in the village shop called me the other day. There is growing local interest in the event, along with growing anticipation of an opportunity for a really good laugh. Competitors are beginning to realise that very soon we will be the focus of the entire village's amused scrutiny and, according to Lollipop, there is much secret practising going on.

We have a group rehearsal on Thursday evenings in the big pub where we will be performing. It has a dance floor and is the obvious, indeed only possible, venue – the punters will need access to drink and plenty of room to roll around laughing. We also meet on Saturday mornings in the village hall. Each couple has a short slot to rehearse their individual routine. Lollipop's dad is the keeper of the keys so she has taken responsibility for opening the hall and setting up a mini sound system for the dancers to practise to. This is very civic minded of her, but she has an ulterior motive. She loiters around and casts a keen eye on the other couples in order to suss out the competition, then she fills me in on who she considers to be our biggest threat. The way she sees it, there are three couples apart from us who she thinks are in contention for the title. There will be three judges on the night, but the audience will be voting too, so it will, in part, be a popularity contest. Seen in this light, there is an obvious winner. The male half of one couple is one of the senior team's recent star players. He is a bit of a hero in the village and is going to garner a huge vote. His dancing style is a little, shall we say, angular, but his partner can throw a few shapes and is well liked. They will score highly. Another couple have an even larger age difference than Lollipop and me. She is an undoubtedly glamorous granny and he is a smiling young fella. Their routine is a slow waltz with a lot of bowing. It is not terribly ambitious but it is very sweet. Lollipop thinks they have a high 'cute' factor. The third couple are, to put it plainly, very good indeed. Generally speaking, the women competitors can all dance well and the men's skills are more variable – but the male half of this last pair can really move. He is young, sporty and can pick up steps in jig time. Bastard. The two of them do a terrifically athletic jive, with lots of throwing and twirling. It's great.

Lollipop and I are doing a Latin/disco routine to 'You Should Be Dancing' by the Bee Gees. We both strut with attitude and salsa with sass. As a finale, she wants us each to do a cartwheel and then I have to catch her, fling her over my right hip then over my left in a kind of scissors movement, before dropping her neatly back on her feet. It will be impressive if it works. She has her eyes fixed on the trophy. As a showstopper, we are all doing a group routine to 'The Time of My Life' from *Dirty Dancing* – inevitably. The instructor has me and Jiving Sportsman doing a climactic knee-slide the length of the floor – just like Patrick Swayze does in the movie. We try to keep it symmetrical and there is absolutely no likelihood that either of us will try to slide just that little bit further than the other. We rehearsed it a few times last night. When I do it well, I get friction burns on my knees and I don't care.

It has been easy to forget that the whole purpose of the night is to raise money for the senior club. We have each been allocated a certain amount of seats to sell and it is tacitly understood by all that, if in doubt, audience members are under a moral obligation to vote for whoever sold them the ticket. This is natural, right and proper. It is essentially the principle which guides the operation of Irish political life. Jiving Sportsman told me, with a sheepish grin, that he had sold three times the number we were originally given. This spurred me into action. I nipped up to the farm this morning to try and shift some of mine. I found Male Farmer Friend in the sheep shed, his arm up to the oxter in the back end of a ewe. Lambing has begun.

Tending Livestock

There will be far fewer lambs up at the farm this spring than there have been in previous years. Maybe sixty-five ewes will give birth compared to two hundred and fifty last March and six hundred or so in peak times. The lambs arrive over the course of about a month – three weeks of uninterrupted birthing, with a few less hectic days at the beginning and end. For the Farmer Friends it is a time of very little sleep – they can often be up all night helping difficult births and grabbing a few hours' rest whenever possible. I remember the look of them during these long, busy days – haggard and grey faced with exhaustion. Over the last few seasons they have been gearing down. They are not getting any younger; their three children have all grown and none are currently living at home. The prospect of hundreds of pregnant ewes and no young people around to help with all-night sessions was, frankly, a little scary, so this year the numbers have been substantially reduced. Lambing is proceeding at a much more manageable pace than I have observed before.

Nonetheless it is still incredibly noisy in the sheep shed. We often walk up there with the boys after school if the weather is fine. The ewes have a constant supply of silage to nibble at but are usually given their second daily dose of dry feed at about four in the afternoon, so if anyone appears in the shed round about that time, the flock breaks into a deafening blare of excited bleating. The first few times I heard this I was astonished by the clamour and surprised to discover that sheep can have quite distinct voices. Some have high-pitched, steady bleats; others baa in short, brassy bursts of baritone; some have evil cackles – like a pantomime villain's manic laugh; and underneath these tones, you can hear an occasional flabby bass chuckle like a jolly fat man shaking his jowls. When the lambs start arriving they add their own piping treble notes which pierce the cacophony like piccolos.

The boys love the din and are always eager to help with the feeding. Big buckets are filled with mixed grain from the large pile kept in the dry barn in the lower yard. These are piled onto the back of the pickup for the short drive to the sheep shed and the boys hitch a ride. Then the grain is poured into the long wooden troughs that form a barrier between the separate pens and the sheep charge and shoulder each other aside to get at it. Suddenly the massed bleating stops and is replaced by general snuffling and chomping. It seems almost peaceful. The boys will then usually jump about for a bit amongst the bales and piles of fresh straw that are strewn down the centre of the shed. When they tire of this they might pick up a lamb to pet. I have a series of photographs taken during lambing each year. In the earliest ones, the boys are grasping lambs nearly as big as themselves. In subsequent ones their stature increases, as does the confidence with which they are holding the creatures. What has remained constant is the look of delight on their faces. The urge to pick up a newborn lamb is a strong one.

Ewes are much more relaxed about people handling their young than cattle are, and very young lambs seem much more inquisitive and adventurous than calves. They can get out of the pens quite easily, and their mothers don't seem to be especially bothered. They skip and frisk about the place with a dainty, high-stepping gait. They often venture outside the shed itself, although they never stray too far, and they always find their way back to the right pen when they are hungry. They feed with as much enthusiasm as they gambol and will try their luck at any available teat that they pass, but the ewe will swat away any lamb that doesn't smell right. When it finds its mother, the lamb will headbutt her udder vigorously before latching on and sucking furiously, twitching its tail like a puppy. Young lambs vary hugely in size – a sturdy single lamb can tower over either one of a pair of twins, and the runt from a litter of three or four can sometimes be only half the size of its biggest sibling. In the case of really scrawny ones, the woolly skin hangs on the body in wrinkled folds until the lamb has grown and filled out.

The flock is part Suffolk and part Lleyn – a Welsh variety – so the lambs can vary in colour. Some are white or creamy yellow, some are speckled and some are black. Milo was the rejected last-born from a litter of three and was a pure Suffolk. He was born black and his face and legs remained so, but his wool grew to white as he got bigger. Donald was a rejected single from a Suffolk dam and a Lleyn sire – he was white all over apart from a little freckling around his eyes. Whatever their size or colour, all lambs have large ears that seem to dwarf their small heads until they have grown a month or two. But not all of them get to grow.

When I called in to sell tickets and Male Farmer Friend was up to his armpit in a sheep, he wasn't delivering a live lamb: he was removing dead twins to save the mother from a potentially lethal infection. He was wearing a double thickness of arm-length plastic gloves, but he expected to reek for days. The smell was like a pungent, putrid cheese. He pulled out the eel-like bodies and left them in the fresh straw so that the mother could grasp what had happened and not be worrying later about where her lambs were. The stink was intense and the sight was gruesome. But these days I am more curious than squeamish.

When Older Boy was still a toddler, we took a stroll up to the farm to show him a cute baby lamb. When we walked into the sheep shed there were plenty about: some trotting around in small gangs and one or two intrepidly exploring by themselves. Some, perhaps only minutes old, still wet and slightly bloody, were asleep in little twitching huddles – nearly invisible in the warm straw.

These were the days of hundreds of sheep and plenty more lambs on the way. Male Farmer Friend was kneeling next to a struggling ewe when he spotted me and said, 'Give me a hand here!' He had me hold the protesting mother steady while he pulled out a stuck lamb before it choked. I was wearing nice cords and a cashmere sweater I had only recently got for Christmas. I spotted a clean patch of straw and reluctantly knelt on it. I laid a restraining hand on the sheep's haunches. An anxious labouring ewe is a strong animal – Male Farmer Friend gave me an equally strong look and I realised wrestling was required. There was no time to nip home to change, not that I would have dared suggest such a thing anyway, so I grabbed hold and pinned her down. Farmer Friend's fist disappeared into the ewe, and then in one slick slither the lamb was pulled out and flopped onto the floor. I let the ewe up and she began to lick her newborn clean. The lamb began to stir and I noticed the tiny, tight curls of its fleece. It lifted its head shakily and gave a tentative bleat. It was the first lamb I had seen born and I forgot the sheep shit on my trousers and the amniotic fluid on my jumper. Nowadays I always dress appropriately when I call up to the farm. Even when we go there for dinner, I throw wellies and an old coat in the car – just in case.

Because of the smaller number of lambs this year, there ought to be a corresponding reduction in the amount of orphans or rejects that will need to be bottle fed, and it is unlikely that Male Farmer Friend will have to perform the skin-a-dead-lamb-to-fool-its-mother trick. But he has skinned a few calves for the same reason and I helped him with one. A calf had died and another one's mother had no milk. Bottle feeding a bull calf is an entirely different proposition from bottle feeding even the hugest of lambs, but the calf needed to suckle and something had to be done. When I walked into the lower yard the dead calf was hung by its front legs from the raised hydraulic arm of the Loadall; the detached earth-moving bucket was on the ground just below. The bereaved mother was in the barn, lowing loudly for her child. In the same pen were the hungry calf along with its mother and her empty udders. Male Farmer Friend was just about to start the macabre but necessary procedure. He cut into the hide to make a line around the neck and along the length of the belly, and then he started to peel back the skin. He cut neat loops around the top of the fore-legs, like stocking tops, and then pulled the hooves through. Next he pulled the hide further down and looped it around the large cast-iron hooks of the Loadall's bucket. Then he slowly lifted the hydraulic arm and the whole hide was pulled free. It was like peeling off a sock inside out. My job was to assist the process by applying a sharp knife here and there to help separate the skin from the flesh when it was necessary. Luckily it wasn't needed that much, as I was pretty ineffectual at this. I didn't quite feel queasy, but it wasn't a pleasant job and I wasn't performing it with relish. The Beloved is bemused by my desire to be involved when Male Farmer Friend is doing such things. I am not entirely sure myself why I ask him to text me if he is up to anything 'interesting'. There is a curiosity, certainly – perhaps a morbid one – but also a sense that I should see it all. I love the sight of cattle and sheep in the fields around me. When the young calves are bucking and the lambs are leaping, it makes a pretty pastoral picture but there is a darker side. Balanced against this proliferation of life, this abundant sense of growth and increase, there is occasional death and subsequent real loss, which I feel I ought to know about. The great thing is to turn the loss into gain. And this is what we did.

Male Farmer Friend got the hungry calf out of the barn and I held him while the hide was fitted. His legs slipped neatly through the loops that had been so dexterously cut, front and back, then a bit of string fastened it at the neck and the job was done. The calf seemed to be wearing a cowhide cardigan with attached leggings or chaps – a kind of leather 'onesie'. It looked slightly ridiculous, but I still remembered the grim preliminary operation, so I didn't laugh. However, I was soon smiling because next came the hugely satisfying conclusion. We led the calf into the barn and pushed it towards the mourning mother. She sniffed it and accepted it at once. Male Farmer Friend stuck a finger in the calf's mouth. The calf sucked hungrily and followed the finger as it was guided towards the full udder – then with a neat flick the calf's head was deftly transferred onto a teat where it guzzled and drank. The next day I got a text saying the hide was off, the calf was feeding well and the new 'mother' loved it.

Farmer Friend, of course, had to dispose of the dead calf in the same manner I described earlier, and I am sure he felt its loss far more deeply than I could. He has lost calves before and will lose calves again and will more than likely end up skinning the odd one for the same excellent reason, but I don't think he has become inured to the grisly nature of that particular task, nor to the general muck and mess his job involves. At one point during the skinning he got a whiff of my new birthday aftershave and he said, 'God! That's a lovely smell! All I seem to smell at the moment is shit and stink.' I also know that every dead animal depresses him – not that there are that many. The vast majority of the herd and the flock thrive. The cattle and sheep that surround me are fine, well-nourished animals, and not just in my ignorant opinion – I have often heard the vet remark on their health and vitality, and he is a man of few unnecessary words. I suppose that because the Farmer Friends are in the business of rearing live animals, the loss of one can be seen as a failure and failure is felt more sharply than success. For most of the last ten years, I have taken for granted the bucolic beauty of the livestock grazing contentedly in the green fields. Gradually I have become aware of the effort, determination and disappointment that real country living entails.

The other morning we ran out of coffee and the Beloved was getting twitchy. The village shop has only basic provisions and all I could find was instant, so I nipped up to the farm on the scrounge. I found Female Farmer Friend in the sheep shed with a ewe whose right udder was full to bursting. She had borne twins: one had died and the survivor was feeding from one side only. Maybe on some dim subconscious level it felt the other side belonged to its deceased sibling and was leaving well alone. The result was that the ewe was in need of extra relief to avoid mastitis and infection. Female Farmer Friend told me she was trying to get other lambs to feed from the swollen udder until the ewe's actual offspring copped on and drank from both. This was the beginning of a lesson in the post-natal care of lambs. I used to think that a day or two after birth they were all out in the fields with their mothers, delighted with themselves, eating grass and chasing other lambs from high ground to low ground and then back again. This is what some do, but others need nurturing before they can leave the sheep shed.

For a start, they can't digest grass until they are about six weeks old. Apparently they have a series of stomachs: the first one digests only milk and at a certain age it withers away. Then the lamb starts to eat grass, which is processed by the second stomach, where it is duly turned to cud, which then proceeds to the third stomach where that, in turn, is dealt with. I don't really know what happens next, and I don't particularly want to. The point is, if a young lamb is out in the field and not getting enough milk, hunger will make it turn to grass, which it is unable to digest because it will stick in the first stomach, which is unable to break it down. The lamb will then get a bloated belly but remain skinny, which is not good, nutritionally or aesthetically. So you have to mind the lambs indoors until such time as you are confident that they are suckling properly.

There are a host of other things to watch for too. Female Farmer Friend went on: an older ewe having borne many lambs will have saggy udders – she paused as we both considered an analogy which neither of us dared mention. The problem this can cause is that the milk drops down and collects below the teat. A smart lamb will buffet and nudge away until the milk comes – I have seen this – but some lambs are stupider than others and the udder has to be manually massaged until there is a proper flow. I have seen this too. In the case of twins or triplets and, when they occur, quadruplets, the smallest will suffer if nature takes her course. The larger ones will use the advantage of size to drink their fill and thus grow even bigger, while the smaller ones can languish and dwindle. So you have to keep them indoors and equally fed until they are even in bulk – only then is it safe to let them out. Sometimes, perversely, size and strength can be a disadvantage: it is not unknown for a ewe to produce a single hefty and healthy lamb and then have no resources left to produce enough milk to feed it.

Female Farmer Friend can spend weeks and weeks after lambing has supposedly finished pummelling udders and massaging teats to feed runts and giants alike. And then there's the fox! It is astonishing that so many lambs survive at all. Of course, in the wild they wouldn't. It is only because of the tireless work of people like the Farmer Friends that there are so many sheep for us all to eat. Hooray! My lesson ended with a round-up of local gossip – according to the current buzz in the village, Lollipop and I, along with Jiving Sportsman and his partner, are joint favourites for the dance competition. Expectations are rising and the big poster in the centre of the village with photos of the competing couples has 'Sold Out' emblazoned across it.

The crocuses are finished. Even the later purple ones have dropped their petals and disappeared until next year. The grape hyacinths under the silver birch are starting to reach up and the narcissi are blooming. These diddy daffodils are one of my favourite flowers. They have the bright cheerfulness of their full-sized cousins but less of the noisy riot. We have a host of golden daffodils – the Beloved has planted Wordsworthian quantities over the years. When I have opened the shutters of the boys' room these last few mornings, they shout at me from clusters around the piggery, and there are hordes of yellow batches interspersed among the fruit trees further down the field. The grass is lush and thick and the buds of the large cherry tree look ready to burst into blossom. In some years all its pale pink flowers open at once in a glorious single blaze. In other years the blossoming is staggered: it flowers branch by branch in a floral Mexican wave. I have no idea why it varies like this.

The vernal equinox is approaching – halfway between Imbolc and Bealtaine (the beginning of the summer), it is the supposed midpoint of spring. It is one of the two days of the year when night and day are of equal length. To the ancient Celtic mind, the equinoctial equilibrium between light and dark represented a magical balance of nature and the cosmos.

There has certainly been a balance in the weather – it has been equally fair and foul. A few days ago it was beautiful. The Beloved and I were working outside in the sunshine. We started to prepare the kitchen garden for planting and a few bumblebees nosed around the hyacinths and anemones at the front of the house. The boys claimed summer had arrived and initially refused to don vests before skipping down the hill to school. We knew better but still had to cajole and coerce before they submitted. Then we had heavy rain, sleet and bitter winds, and it wasn't at all gratifying to be right. The prevailing weather and the local temperature can vary wildly during the course of a single day. I will often light the stove first thing because of the shivery chill in the morning and that same evening we might be sweating in the sitting room.

Whatever the vagaries of the climate in March, as far as my childhood recollection can ascertain it was always traditional to have a cold and wet St Patrick's Day – in Dublin, at least. However, as I remember, it never rained quite enough on the seventeenth to dissuade my dad from dragging my sister and me into town to watch the parade. The day usually began with an enticing suspicion of sunshine so we would head for the city centre, but by the time we found a vantage point on O'Connell Street a steady drizzle would have settled in. We watched bare-legged, dispirited drum majorettes as they led yet another pipe band through the mist. Periodically the pipes fell silent and I would listen as the massed drums beat a tattoo to the rhythm of my chattering teeth. Still, it was a day off school and there was a nebulously festive air. We usually scored an ice-cream, despite the cold, and I was always proud to sport my bit of shamrock or a green, white and orange rosette. But that was in the 1960s – the twenty-first-century celebrations in Dublin have massively improved and the weather hasn't seemed as bad these last few years. The Beloved has an annual gig commenting on the parade for the edification of the invited VIPs, who get to sit in the specially constructed stands along the route. As a result, she gets a very good view of the elaborate floats and the giant puppets and the fire eaters that accompany the traditional bands. But there is only so much you can say about the relative mace-twirling styles of the girls from Massachusetts compared to the girls from Sydney, so she often entertains herself and her stand by departing from the usually dull script and collaring people, microphone in hand, in order to ask unexpected questions.

This year she was on the primary stand with the president himself. (Older Boy remarked that 'Mummy does important things.') She didn't have as much fun as in other years. Our president doesn't seem to be a man who insists on the formalities that attend his office, but his 'people' are sticklers for protocol. At one point the Beloved interviewed the man who drove the motorised 'pooper scooper' which followed a smart unit of mounted police. She asked him how he saw his contribution to the national day. He replied, 'I'm the shit shoveller!' When she stepped back onto the stand, she was primly told by minders 'not to talk to people in the parade'.

The boys and I had more fun in the nearby small town where we usually go. The first year we went, I was surprised to find it so easy to bag a good spot to watch from. The town is small, but its population is in the thousands – I expected at least a little jostling for position to get a decent view. I also recognised most of the faces around us from our own village or other neighbouring ones. This was because the majority of the people of the host town were actually in the parade, and this year was no different. My mum was with us for a few days and she came along. She saw no reason to defer the boys' gratification, so as soon as we staked our claim to a stretch of wall at the bottom of the main street, she bought us all giant green ice-cream cones and we sat back to watch. First came the pipe band: they stopped playing just before they got to us – I think they had run out of tunes – so for us the parade began in silence. This was broken by the loud siren of the local fire engine which followed; then came the nursery school in assorted costumes, waving balloons and blowing whistles; then the scouts and guides leading an awful lot of tractors, one of which appeared to be driven by a gorilla. All were met with appreciative applause.

A slightly louder wave of clapping signalled the arrival of Saint Patrick himself – a tall man with a stupendous beer belly. His costume was a large white nightie and a green-felt mitre. He had a false white beard over his real brown one. Behind him was another tractor, driven by a man with a green, white and orange Mohican. The tractor was towing a nearly life-size plywood airplane with an open cockpit. In the cockpit sat a goat wearing a green Stetson. Many trailers passed carrying various townspeople doing various things. One woman sat at a sewing machine as she waved and nodded. There was a man thumping an anvil with a lump hammer. These were followed by a larger trailer with a mocked-up kitchen in which a woman stood at a cooker and pretended to fry an egg whilst a man sat at a table and fiddled with an old wireless set. Then there came the local amateur dramatic group – my favourite. Frozen in different theatrical poses, they wore vaguely eighteenth-century outfits. The driver was finding it difficult to keep his speed consistent, so they kept wobbling violently as he lurched forward or slowed suddenly. At the back in a rocking chair, which was absolutely not staying still, was an old man in drag. He had a purple wig and a slash of red lipstick and he was draped in a huge shawl and a long black skirt. He had, to quote my grandfather again, a face like a smacked arse. His expression, as he glared at the spectators, suggested that he had probably thought this an amusing idea in the pub but was now seriously regretting it. His eyes met mine and I blew him a kiss. His lips moved furiously and filthily as he looked indignantly away. The rear was brought up by more tractors, a steam engine, a really rusty old combine harvester and a handful of classic cars. Every offering was loudly and good-naturedly cheered by the crowd. The general feeling every year seems to be that, though we all know the parade is an artless, ramshackle hotchpotch, it is simultaneously fabulous. For the boys, there is the added attraction of sweets being scattered from most trailers as they pass, and you don't get free lollipops on the presidential stand.

When I lived in Dublin, Paddy's Day in the city centre was always a little hazardous. Quite apart from those locals who wished to celebrate en masse, the town was awash with tourists in giddy pursuit of 'the craic'. One year I was in a bar frequented by actors. In honour of the day, I had the night off from the show I was in at the time. Paradoxically, most Dublin theatres were open but the Lyric in Belfast, where I was performing, was shut. It is situated in a Protestant area of that divided city and I suspect is seen as a nest of Fenian vipers by the surrounding residents. I enjoy Belfast, with its fierce humour and bloody-minded conviviality, but tensions used to rise on certain red-letter days so I decided to head home to toast Saint Paddy. The pub I was in backs onto an alley, which is also opened onto by the stage door of a venerable old theatre. Half an hour before show time, an actor I know wandered through the bar on his way to do the play – he was masochistically investigating who was out having fun whilst he had to go to work. He was extremely disgruntled and explained why. Red haired and not very tall at all, he had grown a thick beard for his part. He had pulled on whatever had come to hand that morning – as it happened, an Irish rugby shirt. So there he was, a short man with a red beard and wearing green, minding his own business as he walked through Dublin on Saint Patrick's Day. Apparently he had been groped and manhandled by a group of American women. They had picked him up and shrieked, 'It's a leprechaun!' Best stay out of town, really.

My most enjoyable Paddy's Days have been spent out of the country – the day seems to achieve its apotheosis abroad. There is a fiercely joyful pride in the way the Irish diaspora celebrates its nationality. I remember drinking green Guinness in Manhattan one year, which was disconcerting to begin with and hilarious after a few pints. But the most memorable, for sheer exhilaration, was in London, in my twenties, at a Pogues gig in Brixton. Shane MacGowan roared and growled on just the right side of lucidity and the band rocked with tightly controlled chaos. They were all wearing the uniforms of New York cops, which seemed to declare that you didn't have to be Irish to be Irish, but if you were actually Irish, and expatriate Irish at that, then you were even more so. The music turned an already happy crowd into a delirious, seething mass. We jumped and jigged and leapt and sang and shouted and swayed all night. I weaved home in the small hours drenched in beer and sweat. I haven't been to such a concert for decades – I suppose gigs can only be like that when you are young.

The fact that the day is associated with copious drinking and eating has, in the past, caused problems – for the Catholic Irish at any rate – because it falls smack in the middle of Lent, the forty days of abstinence before Easter. But an Irish solution was found to an Irish problem: a tale was concocted to demonstrate how the saint himself 'relaxed' his own Lenten fast for a day. Eating meat was prohibited but fish was allowed, and the story goes that Patrick couldn't resist temptation and ate some meat during Lent. Being very holy, he then repented and prayed. An angel appeared and told him to put more meat into boiling water; he did so and, lo, there was a miracle! The meat had become fish and Patrick wolfed it. So thereafter, the Irish ate as much meat as they wanted on the national holiday and called their dinner 'Saint Patrick's fish'. The church surrendered to an inexorable force and duly exempted the Irish people from Lent for the day.

My mum expects to be extremely well fed when she comes to stay with us, and when the Beloved returned from the city, I had our Saint Patrick's fish ready. We sat down to the last brace of pheasants from the freezer. I casseroled them with apples, onions, potatoes, some good local pork sausages, bay leaves from our bush and a generous slug of cognac. Whilst rummaging in the freezer I noticed a slight depletion – we still have plenty of lamb but the game is nearly finished. I can shoot more rabbits and pigeons but there will be no more pheasant till they are back in season next November, and the stock of frozen vegetables is nearly gone. It really is time to start planting.

It is traditional to plant potatoes on Saint Patrick's Day and this year I was only a little late. On a dry morning a few days after Mum had returned to England, I was in the vegetable garden digging. After such a wet February, the ground was soft and digging the drills for the spuds was easier than I expected. Perhaps I am getting stronger? Male Farmer Friend dropped down a large load of cow manure and I laid plenty of this in the trenches, with a light covering of topsoil and a line of seed potatoes on top, before covering the lot with earth. I always plant two early varieties – one waxy and one floury. We both love Wexford Queens – they are unutterably tasty when freshly dug, then steamed until slightly fluffy and slathered in butter. So we have a drill of these each summer. For salads this year I have planted Home Guard; we usually sow Jerseys Royals or Charlottes but I couldn't find any of those. Home Guards are reputedly good and waxy so I am trying them out – we shall see in June. I also planted broad beans, parsnips, shallots, a first sowing of carrots and peas and I finally got the garlic in!

The March weather has been mixed to the end. Sun has been followed by snow. Mild days have been laced with raw winds. There have been clear blue skies with scudding white clouds and there has been rain billowing from heavy grey cumulus. All of the separate armies of daffodils are now out and the big cherry tree is beginning to blossom evenly but slowly. It seems we will have a gentle unfurling of flowers this year – a simultaneous fade-up of colour throughout the whole tree. Spring is gaining strength. The Divine Hag is falling asleep and Brigit is in the ascendant.

So I filled in the questionnaire for the dance-off in a more optimistic frame of mind. At the Beloved's suggestion, I am calling myself Sugar Daddy. As I wrote in the name her eyes danced with delight. I have no doubt that she loves me but she also adores a laugh at my expense. She can snigger all she likes – the latest word is that we have emerged as clear favourites. Lollipop is very excited. She has got herself a terrific outfit – skin-tight black trousers flared at the knee and a psychedelic top. I have heard rumours that Jiving Sportsman has acquired a scarlet ruffled shirt with billowing sleeves – I am going to have to scour charity shops for something equally lurid.

Dancing Farmers

The roads are busy with agricultural machinery again, but now the tractors are pulling implements associated with the spring: ploughs, harrows, seed drills and muck spreaders. The Farmer Friends usually let us know when they are using the latter, as the smell can be powerful and pervasive. They suck up the contents of the slurry pit by the side of the cow shed, which can amount to thousands of gallons of accumulated excrement. This is then spread liberally over the fields which are to be planted. It is a most effective fertiliser. I don't object to the smell any more – in fact, I find it almost pleasant. It now has associations of warmer weather and the sowing of seed, which reminds and encourages me to plant my own small plot. Besides, this year the aroma of muck spreading is only a vague and distant hint on the wind. The Farmer Friends will not be ploughing, spreading or sowing the field opposite us because it is being kept as pasturage for the herd and the flock. Up at our end, sheep are grazing and lambs are frisking. Further down in the valley, cattle are drowsing and calves are trotting around in gangs.

Sliabh Buí is changing colour. The buds are opening in the deciduous woodlands on its eastern flanks and the fields on the western slopes have been ploughed. What was green is now brown and what was brown is slowly turning green. Larkin describes trees coming into leaf as like 'something almost being said'. This is a lovely notion, but it suggests a muted, gradual transition. There has been nothing muted about spring's arrival over the last few days, especially in the mornings. Since the clocks went forward and it has begun to brighten earlier, the dawn chorus has been a full-throated shout and many avian calls continue throughout the day. Squabbling male blackbirds fight to establish their territory – the woodpile has been a fiercely contested battleground. Robins do the same in the fruit trees. The hedges are full of the tweets and chirps of various little brown birds. And high above the strife, pigeons coo in the leylandii trees and rooks caw from the beeches. Beneath the birdsong I hear the steady gurgling murmur of the stream.

The crescendo of sound is matched by burgeoning colour. The April sun is still low in the sky but is nonetheless bright for all that – it casts long shadows on the rich grass thick with new growth. The fields stretching away to the distant hills are lined with dappled hedges of bright yellow gorse, and here on our own acre the big cherry tree is in glorious full blossom. After the recent dreary weather its first flowering of bright white almost hurt the eye, but during the last week a hint of pink has been slowly asserting itself – like that film effect where colour is bled into a black-and-white still just before it comes to life. Fat, furry bumblebees, in black-and-yellow blobs, nose around the plum-coloured stamens at the centre of each flower. In a light breeze the branches dance but the blossom stays fixed like glue and the tree looks like a whirl of static confetti. This morning I saw another change in hue – the blossom is now slightly tinged with green as the leaves behind the flowers start to open.

Even the evergreen hollies across the road, the ones that frame our view of the yellow mountain, appear to be greener than they have been throughout the winter. Mauve tulips are blooming in pots at the front of the house and ants are starting to explore between them. The daffodils are turning brown as they begin to droop and wilt, but the smaller cherry tree – the one under our bedroom window – is starting to flower. This one has a pure white blossom and its branches spread upwards and outwards in perfect Zen-like proportions. It could be in a manicured garden in sight of Mount Fuji, except that the grass around it is long, hairy and really needs mowing.

The weather is improving. There is still occasionally heavy rain, naturally – inevitably – but it is growing warmer. We are not habitually lighting the stove every morning. This weekend we went for a walk along a muddy lane flooded with puddles and the boys charged and splashed through them in wellies, shorts and T-shirts. They got completely drenched and squalidly muddy but nobody cared except Perkins, who had to deal with the consequent laundry. The Beloved has taken to wearing shades again; the other evening we had an aperitif outside for the first time this year; and the broad beans and shallots I planted in March are starting to come up. This promise of summer is much discussed in the village. Like anywhere else, seasonal change is a socially useful topic of conversation, but in a largely agricultural community like ours, it is also a matter of vital, material importance. It affects lambing, calving, ploughing, sowing, harvesting and haymaking. However, for the last fortnight or so, before and after the event, the weather has been supplanted as the chief subject for discussion. In the shop, in the pub, at the school, on the road – every conversation I have had has turned inexorably to the dancing competition.

The anticipation had been keen and for the last few days before the big night people without tickets had to call in favours to ensure getting in. The gig was packed. Burly players from the senior football team acted as security to prevent overcrowding by unruly gatecrashers. They doubled as minders to escort the dancing couples through the boisterous crowd to the dance floor.

I had finally learnt our routine and was confident that on the night I would gear it up a notch or two and really sell it. Lollipop's cartwheel had been getting ever more athletic during our last few rehearsals and she had great plans for 'big hair'. I found a garish purple shirt with stripes of many colours – some of which even teamed with her technicoloured blouse. I borrowed Younger Boy's star-shaped shades and purchased a seventies-style stick-on handlebar moustache. I thought this would clinch it.

The running order had been decided by the drawing of lots and we were to be fifth out of nine and would be closing the first half. I thought this was perfect. Too early and the audience wouldn't be warmed up, but too late and they would be probably drunk and possibly rowdy. Glamorous Granny and Young Fella drew first place but they felt their slow waltz wasn't punchy enough to kick off the evening, so they asked to switch positions with a couple who just wanted to get it over with. Everyone agreed so the running order was amended accordingly. At the same time, another couple asked to swap places. No clear reason was given but, as the dancers they wished to switch with were amenable, nobody objected – what did it matter who went last?

The whole evening was being filmed and simultaneously relayed to the large TV screens dotted about the pub – the ones normally used to transmit sporting events for customers to cheer or boo at, so I suppose it was appropriate that we should be thus broadcast. For a dressing room, we were all piled into the small and poky pool room at the opposite end of the main bar from the dance floor, and someone had rigged up a live feed to a laptop to serve as a show monitor – this would allow us to keep an eye on proceedings in order to be good and ready for our spots. When the first dancers were led through the excited crowd, we huddled around to watch intently – if we were going to be jeered or laughed at, this would be the beginning.

But that first couple hit the ground running. She was wearing a short blue sequinned flapper dress and looked stunning. He was wearing a matching blue tie with a black shirt and black trousers that appeared to have trouble staying up properly. He loves a salted peanut and is quite short and quite round. During rehearsals he had seemed to have even more difficulty remembering his steps than I had, and we were all a little worried that he might not get through it. Consequently, our vocal encouragement was proportionally louder for him than for any other dancer. Well, Peanut Man nailed it. Mostly. He hardly missed a move. During their number, he kept his eyes fixed firmly on his partner and followed her as best he could, even if he was therefore just that little bit behind. The only time he looked away was when he mimed shooting a six-gun at the audience – I am not quite sure why. His partner beamed joyously and the audience whooped their approval. Once or twice we heard loud male roars of 'Good Man!' and 'Go on, yeh boy yeh!' It was a brilliant start and we all jumped up and down and hugged each other with delighted relief. Later, during the interval, I went to the loo and overheard deep rumbling voices echoing from the urinals. Two big men were ponderously discussing something important – the fall in the price of beef, perhaps, or the dangerous possibility of drought again this coming summer. I tuned in to what they were saying and realised they were expressing admiration for Peanut Man and the bravery it took to go first: 'Sound man!' said one. 'He did well so!' replied the other.

From then on, everyone stepped up and stepped out. Every routine went the best it ever had, and the audience grew louder and more appreciative with each dance. Lollipop and I took to the floor like hungry wolves and danced like oiled tigers. The adrenalin jolt was the biggest rush I have had on stage in years. Halfway through our routine, which seemed to be flying by lightning quick, I could feel my moustache loosening on my sweaty upper lip, but it stayed on. We raised our game many notches and we knew it. We left the floor and made our way back through the cheering crowd as they shouted superlatives and shook our hands or gave us hugs. Someone gave me a congratulatory thump on the back and my moustache flew off. I didn't care – it had done its job and I wouldn't need it for the group routine … But hang on, I thought, what about the reprise of the winning couple's dance? Would I not need it then? Ah, hubris … it was not to be.

The second half of the show began gently with the lovely slow waltz from Glamorous Granny and the Young Fella. The audience adored it. I was watching from out front and I swear I saw tears in some eyes. Jiving Sportsman and his partner executed their routine with sharp aplomb and his shirt was perfect, but I was quietly confident – until the last couple took to the floor. I can only call these the dark horses. During rehearsals their number had seemed fun but low key, with apparently simple choreography. They undoubtedly had the best tune – Ike and Tina Turner's version of 'Proud Mary' – which was sure to get the crowd going, but Lollipop had never mentioned them as a threat and I hadn't been paying attention. At a superficial glance, they might appear mismatched: she is a strong, generously proportioned country girl and he is a scrawny slip of a boy. They are both great fun to be around and they seemed to be in it purely for the laughs – at one point in their dance she bodily flung him across the floor, which made me think they were going solely for comedy. I didn't see the signs. I had noticed that they were very light on their feet, but I hadn't really registered the implications of the fact that he was the person most people went to when they couldn't quite remember a particular step. Nor had I thought much about how she appeared to have a clear grasp of the group number without ever seeming to pay attention. I have since discovered that he has been tap dancing since the age of three and she has been a frequent entrant and regular winner in Irish dancing competitions. On the night, the Dark Horses rose to stellar heights. Their routine was a joyous, infectious, perfectly symmetric delight and the crowd went wild. I began to have my doubts about grasping the trophy, but there was no time to ponder – we had to change for the finale and then there was the raffle to be held while the judging got under way.

The Beloved had been roped into being MC for the night because word had got around about her Paddy's Day commentary for the president and this was seen as adequate qualification. She announced 'Nine Patrick Swayzes and nine Jennifer Greys' as we made our entrance for our *Dirty Dancing* ensemble number. The women were in identical fifties-style dresses, each in a different colour, and the men wore black shirts and trousers with coloured belts to match their partners. We looked good. The crowd weren't expecting this added bonus and they seemed overjoyed to see us all again, or perhaps they were pissed. The exhilaration was general and, relieved to have got through our individual routines so successfully, we collectively felt relaxed and happy and it showed. We had all taken a few jars since finishing our respective numbers – some more than others, and I admit to being one of the some – but despite this, it was tightly synchronised and uniformly elegant. Jiving Sportsman and I both got friction burns from our knee slides – the delighted whoops from the audience drowned out any yelps of pain. We took repeated bows en masse and headed back to the bar to await the verdict. This took an age. The three judges retired to compare notes and the raffle began. Every raffle ticket sold came with a piece of red card which was to act as a single ballot. People could buy as many tickets and therefore as many votes as they liked, and each vote was to be deposited in one of nine boxes on a trestle table at the edge of the dance floor – each box corresponding to one of the competing couples.

It took time for the ticket sellers to traverse the room and it took time for the electorate to consider their selections. The Beloved had to keep the commentary running throughout the raffle and the voting, and at one point she was distinctly heard to mutter, over the mike, that she had lost the will to live. I popped into our pool-room dressing area to change back into my disco shirt and noticed a group of competitors looking at the laptop. They were intently watching the empty dance floor. Then I realised they were staring at the ballot boxes and making careful note of who was voting for whom. Eventually, the count was complete and we trooped back on to the small stage at the back of the dance floor for the announcement of the best dancers. It was the Dark Horses, of course, and they were very worthy winners … But, I admit, I was disappointed. In fact, I was surprised to discover that I was more disappointed than I remember being when I was nominated for, but didn't win, the award for Best Actor in our national theatre awards a couple of years ago. This suggests some sort of reordering of my priorities – for good or bad, I am not sure which. I was especially disappointed for Lollipop – I would have been very pleased for her if we had won. But we will both get over it and, besides, the night was a memorable triumph for everybody concerned: the show was a tremendous success, and we raised a shedload of money for the club. And even though there have been dark mutterings of 'you were robbed' from amongst my core fan base, I have ignored these. Got a Few Bob Friend, in particular, has hinted at underhand manipulation of the running order for the benefit of the winning couple. He may very well say so – I couldn't possibly comment.

We held the competition on the last weekend before the school's Easter break. The last day of term was a half day and there was a jubilant atmosphere in the school playground as the children came hurtling out at half twelve. It took longer to round them up than usual – they careered about the place as they noisily celebrated their fortnight's freedom and seemed strangely reluctant to leave the school grounds. Younger Boy demonstrated his latest figure-scooting move: I call it the Gliding Buddha. He gets a good speed up, then stands on one leg whilst draping his other leg over the handlebars. Then he closes his eyes and opens his arms slightly, elbows bent, forefingers and thumbs pinched together, other fingers spread like a meditating Yogi. He floats serenely for about twenty feet and opens his eyes just before hitting the wall, then swerves in a long, smooth arc around the basketball hoop. It's his best yet.

The first day of the holidays was beautiful. The boys threw on T-shirts and shorts – with no vests! Older boy said, 'I haven't felt so loose and free for ages!' We went down to the river with flip-flops and fishing nets. They spent a happy hour or so hunting for minnows and I did the crossword in the sunshine.

Everything is warming up nicely. The firewood is nearly gone, but I think we will make it. Blossom is falling from the big cherry tree – in a light breeze the pale petals flutter to the grass like a flurry of unseasonal snow. However, the smaller cherry under our window is still in full flower. This morning whilst planting peas, I glanced up and was dazzled by its crisp white blossom against the clear blue sky. It was like a sharp breath of clean air. Birdsong is constant now. Magpies are chattering and an occasional pheasant hoots. A lone, victorious blackbird seems to have won the territorial struggle and is exultantly proclaiming his supremacy in song. Each morning he sings from his perch on the wire that droops from the old telegraph pole, thick with ivy, at the front of the house. He is announcing his lordship of the domain in the hope that a passing female will join him. The other day I saw a single swallow making its first aerial reconnaissance. It swooped and zipped around the house a few times before disappearing again. There should be a few back by the end of April; for the last few years we have had two nesting couples spend the summer in our garage.

The football season has restarted with the waxing of spring, and one fine evening I drove Older Boy to his first fixture in the new county grounds half an hour to the north of here. The playing fields are on a high plateau surrounded by blue hills and open sky. It must be bleak in bad weather: wet, windswept and grim. But on a bright, warm evening in April it makes a lovely vantage point from which to see golden sunlight pick out the fat, buttery cows in the distant meadows. The team looked good and played well in their smart new kit: their level of skill has stepped up to match their pristine outfits. They won all three of their short matches. Officially, no score is kept, but the parents, as usual, kept a precise tally of every point – along with shouting encouragement, of course. At one point I heard a broad Wicklow accent urge Chantelle to pass the ball to Yasmin. These are not names I associated with the sport when I played it briefly forty-five or so years back. Watching Older Boy acquit himself well as goalie for one match and then loiter around the opposing team's goal for another brought back memories of my own playing style all that time ago. I used to avoid the hurly burly of midfield and linger far forward in the hope of a quick pass for an easy score, thus achieving maximum glory with minimum effort. This tactic occasionally worked, but it didn't survive transposition to England. When I tried to do the same in soccer, people bellowed 'Offside!' at me. I had no idea what they meant and I suspect that sometimes my accusers weren't entirely sure either. Anyhow, my limited Gaelic skills didn't travel well and my soccer career was as brief as it was confusing.

The boy seems to be a better player than I was. He certainly enjoys it, and he was cheerfully chatty on the return journey. The sun was slowly setting as we headed home. On one westward stretch of the road, we seemed to be driving straight into the glowing heart of it. Oscar Wilde loved a paradox and he used to talk about how nature imitates art. He suggested that the actually occurring beauties of the real world were only striving to equal the perfect, idealised images evoked by great painters and great writers. As we drove along the quiet country road, I felt we were in the foreground of an eighteenth-century painting of Arcadia. The light and landscape mirrored each other in a moment of brief harmony which I can still see with my inward eye. The pale-pink blossom of cherry trees, the white blossom of blackthorns and the saffron-yellow gorse reflected the pink, orange and yellow sky. Away from the centre of the sunset, the sky became pale blue then darker blue – which in turn blended with the blue, green and grey hills. Dusk was gathering but the whole effect glimmered for an instant as the light dwindled – like an opalescent oyster shell which is simultaneously silver grey and also a kaleidoscope. I remarked on the beauty of the sight to the boy in the back, pointing out the colours as I saw them. He didn't say a lot – I didn't expect him to – but I glanced in the rearview mirror and saw him look, listen and nod. The last part of the drive was in silence. It was still just twilight when we got back to the house, but quiet no longer. Bulls were bellowing in the fields behind us, lambs were bleating in the field opposite, distant dogs were barking to each other, somewhere a tardy cock was crowing and the stream was gushing noisily.

Hiding Chocolate

The village is still discussing the dance competition and the Beloved continues to be congratulated on her performance as MC. I was a little too absorbed to pay much attention at the time and the sound system in the pub didn't really penetrate to the pool room where I spent much of the show, but I heard her make the occasional witty remark and she kept things moving with brisk efficiency. She has made a great impression and her standing in the village has altered slightly. She now has a reputation for being a bit of a wag, especially in the village shop. When she is chatting to the girls behind the counter they are apt to start giggling prematurely in anticipation of an amusing sally. And when she does actually make a humorous remark she gets an enormous laugh. Seemingly she is great craic altogether. Some of my fellow competitors, already feeling nostalgic, are keen to relive our night of shared glory and I am frequently asked if I miss the dancing and if I regret not bringing home the trophy. I don't, although I would have liked to have won a big cake, which is what dancers used to compete for in April.

There used to be a festive occasion called the Cake Dance. Held on Easter morning at a crossroads, it was a rite of spring as well as a celebration of the end of the Lenten fast. A large cake decorated with spring flowers was awarded to the likeliest young couple, although it was expected that they then share the cake with the assembled company. The custom has long since died out but it lingers in the language: it is thought to be the origin of the phrase 'that takes the cake'. A more pious association of dancing with Easter was the belief that at dawn on Easter Sunday morning the sun danced with joy at the saviour's resurrection. It was never a good idea to look at the Dance of the Sun with the naked eye, so the reflected image of the dawn was observed in a strategically placed tub of water. A discreet nudge applied by an adult would create a ripple to ensure that the sun did actually appear to dance. It was done to prevent the children's disappointment, or to assist in their indoctrination – depending on your point of view. The ritual's connection of the rising sun with the risen Christ is a perfect metaphor for the origins of Easter – the Christian and pagan elements are inextricably combined.

The death and resurrection of a god is an ancient and widespread myth: the Greek god Dionysus was killed by Titans but brought back to life by his grandmother Rhea; the Egyptian Osiris was drowned in the Nile by his brother Set and then arose due to the devotion of his sister Isis; and the Babylonian Tammuz suffered death and was reborn annually with the solar cycle. Western Christianity still determines the date of Easter according to our circuit of the sun and the phases of the moon. Celebrations of the equinox have been recorded as early as two and a half thousand years ago in the ancient Mesopotamian city of Ur. The Jewish people during their enslavement in that region would have witnessed these ceremonies and were perhaps influenced by them. The setting of the date of the Jewish festival of Passover was formalised after this Babylonian Captivity: Pesach, as it is more properly called, begins with the dusk following the first full moon after the equinox. Ceremonies frequently evolve out of earlier ones and modify them. Jesus travelled to Jerusalem for Passover, was subsequently crucified and rose again – he went to observe one religious ritual and initiated another.

Christ's rising from the dead on the third day was the definitive demonstration of his divinity, and bearing witness to the resurrection is the central tenet of Christianity. I can't help but notice that in the gospels the first person to whom he revealed his risen self, the person who first witnessed his Godhead, was Mary Magdalene: she was the first to encounter both the human and divine natures of Jesus of Nazareth. I would have thought that any subsequent religion based on his words and actions would accordingly honour the centrality of women, but I count magpies and see omens in shooting stars – what do I know?

Easter as it is celebrated today has many other elements surviving from ancient sources: the first full moon after the equinox was also the festival of an Anglo-Saxon goddess called Eostre, from whose name the word 'Easter' derives. She was associated with the moon and represented rejuvenation: her emblems, the hare and the egg, symbolised fertility and plenty. She must be the original Easter Bunny! When Older Boy initially expressed doubts about whether such a creature could be real,

I remember saying I had never actually seen him, but that I had no other explanation for where all the eggs came from. Maybe I should have said, '*He* doesn't exist, but *she* does.' Anyway, the boy knows how to look out for his own best interests and he hasn't questioned a bountiful source of chocolate too rigorously.

On Easter Saturday I gave the grass its first cut of the year. I borrowed a sit-on mower from our mechanically minded neighbour and spend a meditative hour tootling up and down the length of the field. I had perhaps left it a little late this year and the grass had grown thick and tall. I had to take it very slowly, avoiding the marshy corner altogether – the still soggy ground would have been turned to mud and the blades would have been choked by the high growth. It was a satisfying task. The rich smell of freshly mown grass undercut by the sharp whiff of diesel instantly evoked summer, and the sun shone warmly. A whisper of a breeze blew the last petals from the big cherry tree around my shoulders as I passed. No matter – there were still plenty of blossoms. The plum tree was coming out in tiny flowers – white with a hint of yellow – and the fruit-bearing cherries were starting to open. The lush green grass was studded with dandelions and daisies – it almost seemed a pity to cut it. But it would only have grown to seed if I hadn't and wouldn't be nice to walk through. Besides, once mowed, the field has a pleasing pattern of broad, alternating light and darker green stripes – easy on the eye and perfect for boys to run over. At one point I stopped for a break to survey my progress. I shut down the noisy engine and slipped off the ear protectors I normally use for shooting. The sudden, apparent silence was only temporary and I became aware of all the sounds around me: birdsong chirruping from the hedgerows; the lazy, fat buzzing of bumblebees in the fruit trees; and high above, in the clear air, the melodious song of an invisible lark.

The boys were at the bench, shirtless and studious in the sunshine, thoroughly engrossed in some art work. When I finished mowing they showed me what they had been at. It was a detailed map of our acre, marked with all the spots where, historically, the Easter Bunny has been known to leave eggs. Early the next morning, we did our best to ensure the map was largely accurate, although one previously fruitful location is now out of commission. We use to leave a few foil-wrapped eggs in the hen house beside the freshly laid hens' ones. I remember Younger Boy grinning with delight as he reached for the glinting coloured eggs nestling in the straw amongst that morning's clutch … with Goldie looking quizzically on. Leaving anything in the sad, empty hen house would have been cheerless, so to compensate we added an extra location this year – Lesley the Lumberjack's wood pile. We laid a few chocolate coins and rabbits amongst the logs but they were ignored. The spot was not marked on the map and, consequently, the boys ran obliviously by. We coughed and nodded meaningfully to draw attention to the hoard. The Beloved glimpsed Older Boy looking suspiciously at us as we 'accidentally' gestured at the Bunny's new hiding place. Perhaps the game is up as far as he is concerned, but if so he kept quiet and was rewarded with a bonanza of chocolate eggs.

The Easter feast used to begin, not with chocolate, but with real eggs. They were forbidden, along with meat, under strict observance of the Lenten fast – hence the custom of using them all up in pancakes on Shrove Tuesday. Naturally, the hens continued laying eggs regardless and a glut accumulated. Eating as many as possible at Easter became almost a competitive sport. They were boiled in water with onion skins and other natural dyes to pretty them up and make them more attractive to children. The boys were outraged at the suggestion we reinstitute the practice this year – it was hard enough to halt the chocolate binge long enough to get a bit of roast lamb into them.

The notion of eating a paschal lamb for Easter confused me when I was a boy. John the Baptist famously cried, 'Behold the Lamb of God – he will wash away the sins of the world!' As far as I understood it, the paschal lamb and the Lamb of God were interchangeable terms for Jesus. Why were we eating our Lord on the day of his resurrection? St John, of course, foresaw Jesus sacrificing himself to save humanity and was referring to the Jewish concept of the paschal lamb – the offering made to God in thanks for the escape from pharaoh and the flight from Egypt. In ancient Jewish tradition, the sacrificial lamb was killed on the eve of Passover and eaten the following day with bitter herbs and unleavened bread.

This Easter we ate one of Milo's back legs with roast potatoes and glazed carrots. The vegetables were shop bought – only the scrawny dregs of last year's carrots remain in the ground, but there are a few signs of new life. The shallots and broad beans are looking perky. The first early spuds, the Home Guards, are coming up and there is a hint of stiff greenery in the drill of second earlies – the Queens. Very few of the garlic cloves I planted have germinated. Maybe I was just too late. Meanwhile, in the orchard the plum-tree blossom is falling, the fruiting cherries are bursting with bright colour and the pear trees are beginning to flower. I have a lot of planting to do over the next few weeks and another big pile of cow manure to deal with. Male Farmer Friend dropped down a large load in exchange for taking his aunt to the hairdresser's and I should really get it spread before the apple blossom comes and goes.

The kitchen garden is not what it used to be: the planks that define the four beds are mouldering and crumbling; the sand-covered matting between the beds encourages weeds rather than prevents them; and after ten years of planting, the soil in the beds must be nearing exhaustion. There has also been some earth creep. Apart from the hundred and fifty square yards between the front door and the road, there isn't a patch of flat land anywhere on our acre. Tucked as it is behind the hedge just below the top corner of the field, the kitchen garden is the most nearly level stretch we have. But it is still gently sloping, and over the years there has been a slight but perceptible flow of soil downhill. The rotting planks are tilting and warped. The flagstones I laid as paths between the beds are undulating and slightly askew. And in the beds themselves, the soil inexorably gathers in banks along the lower borders and has to be redistributed before it flows over the edges of the wooden boards. It all bears little geometric relation to the complex lattice of coloured twine I staked out when I first made my plans. It would need much work to rectify it properly and I doubt I will ever do it. It took a lot of work, in stages, to organise it over our first years here, and I don't think I could face digging it all up, levelling the land and starting again from scratch. All I hope to manage is the odd patch job here and there, along with regular application of our own compost and Farmer Friend's manure.

I have sweated, cursed and wept in these vegetable beds. The soil is the stoniest in Wicklow – I have blunted my fingers and torn my nails more times than I can remember. Once I badly pulled my back whilst trying to shift a bastard of a boulder and had to be stretchered out to an osteopath. I must have spent months of my life weeding. In the early years, I weeded regularly and intensively as I prepared and dug each of the ten-foot by twenty-foot beds. I gradually learnt to dig deep for the roots of perennial weeds like dock leaves, dandelions and grasses, otherwise they just returned: vigorously, infuriatingly and relentlessly. I remember the year I started to win. The weeds were at bay and I acquired a big load of horse manure to fertilise the whole plot. The manure turned out to be insufficiently rotted and was full of undigested seed. It was like sowing turbo-charged grass in every bed. The consequent weeding was gruelling and perpetual. I cried often. I eventually cleared the supergrass and then a nearby plot of land was left untended for a few years. It became derelict. Ragwort, dandelions, nettles and thistles grew like beanstalks. In the slightest breeze, I could actually see clouds of weed seed billow over the fuchsia hedge to reinfect my beds. I wonder the Ancient Greeks didn't come up with something like this as a torment in Hades. In between Sisyphus forever rolling his rock up the hill and Tantalus unable ever to reach the fruit or drink the water, there should be a Minoan melon farmer cursed to ply his trowel perpetually for stealing weed-killer from the gods.

I wouldn't have persevered if I hadn't had an initial success. In our first spring I cleared a bit of ground, dug it over and planted a packet each of onion, carrot and lettuce seeds. The onions never happened, the carrots were measly but the lettuce was profuse. It grew and grew and thickened and spread. I didn't know about 'thinning out' – the process by which you remove some seedlings to allow others to grow properly – and I ended up with a shrubbery full of lettuce bushes. I was thankful it was behind the hedge – any passing agricultural types wouldn't see the ridiculously large crop and snigger at the eejit who plainly hadn't read the seed packet and had planted, in one go, enough to alleviate a small famine. Still, I was delighted with my success and astonished at how easy it had been: I just threw some seeds in the ground and – lo and behold – they grew into things, things that you could eat … Well, a portion of it at any rate. Any visitor that first summer left with a sack full of leaves.

I began to imagine a horticultural demon drug pusher who gives you the first hit for free and then demands a price. I called him Mulch, and he knew exactly how much to give before making me pay. The following year I had success with broad beans and potatoes but my tomatoes rotted and the carrots and scallions failed entirely. I bought a book called something like 'Gardening for Gobshites' and gave myself a proper scare. When it comes to disease, Mulch's hordes are legion. The list of bugs, mites and viruses that vegetables are subject to is long and horrifying. I have seen many of the depredations described. White rot in onions is a furry fungal growth that destroys the crop and also renders the ground unfit for other alliums (leeks, shallots, garlic and so on) for eight years. Carrot fly can eat the young shoots overnight and kill the plants completely. The section on potatoes consists of two pages on their planting and care followed by three pages on the infections that destroy them – they can even contract gangrene, for Christ's sake! And then there's blight: the airborne virus that notoriously led to the starvation of millions of Irish in the nineteenth century.

It flourishes in humid weather – a warm and wet July or August is perfect. Once it has taken hold, the plants are doomed, and usually once you see the signs it is too late. If vigilant, you might spot the white powdery patches on the underside of the leaves, and then you can cut the green haulms and save the tubers beneath. I wasn't always vigilant. One damp August, my proud and vibrant greenery started to wilt and rot. I dug up the plants to see what was happening and every potato turned to sweet-stinking mush in my hand. What those people in the 1840s must have felt when they saw the same thing happening to their sole source of food doesn't bear thinking about.

One of Mulch's less sinister but nonetheless provoking minions is Beelzebunny. The rabbit population around here rises and falls. In our third or fourth year their numbers boomed and they ate everything. Older Boy was two or three at the time and fond of *Peter Rabbit* as a bedtime story. I was completely on the side of Mr McGregor, the gardener – when his wife cooked Peter's daddy in a pie, I figured he had it coming. I trained the boy to chant, 'Bunnies must die!' which upset his grandmother but gratified me. Sadly he grew out of it and stopped chanting the mantra on demand. It is a poignant thing when they get older and less malleable.

Despite these frustrations and disappointments, the pleasure of growing my own vegetables has far outweighed the pain. Every time I dig a drill of spuds, I feel a glad surprise when the tubers spill from the fork like mined nuggets. There is nothing like the exploding flavour of a tomato picked straight from the vine on a warm summer evening when you are cutting courgettes for dinner. A grin splits my face whenever I picture Younger Boy gnawing a freshly picked and cooked corncob with melted butter slathered on his face. One wet, dreary summer I was standing amongst the tilted bean canes picking peas. Older Boy was three and was with me. It was drizzling and there seemed to be no break in the mournful weather. I handed him an opened pea pod, then another, then another. He gorged himself, pausing only to say, 'More, please, Daddy.' And of course, the Beloved has shared every meal with me. She often refers to 'your vegetable garden', ascribing ownership and thereby responsibility to me. But how do you define ownership? Who is the garden actually for? I certainly do most of the work, but *cui bono?* Who benefits? I like parsnips and she doesn't: I grow one small row of these. She loves beans and I am not so keen: this year I am planting five varieties. I am merely making an observation. We are now in the heart of the rowl of planting time. I have just sown more peas, broad beans and carrots and a drill each of purple-topped turnips and beetroots. As May approaches I will spread and dig in the manure and compost and then start thinking about courgettes, cucumbers, tomatoes and sweet corn.

Yesterday was the last day of the Easter holidays. It was as glorious as the first and we were all outside. The Beloved was tidying up around the roots of the fruit trees and the boys had purloined my hand saw and were cutting small gaps in the large hedge to make hideouts. I was on a step ladder tying back a strand of the climbing rose that snakes up the side of the piggery. I looked around. The two pear trees are in blossom, the three apple trees are showing a hint of pink flower and four swallows are back. Sumer is icumen in.

THE BELOVED'S CHOCOLATE & ORANGE CELEBRATION CAKE

Ingredients

For the cake:

175g unsalted butter, softened, plus more for the tin

100g chocolate (50–60% cocoa), finely chopped

120ml just-boiled water

200g plain flour

1 tsp baking powder

1 tsp bicarbonate of soda

100g ground almonds

275g dark muscovado sugar

zest of 2 oranges

1 tsp vanilla extract

3 eggs, lightly beaten

150ml buttermilk

For the icing:

90g chocolate (50–60% cocoa), broken into pieces

40g unsalted butter, softened and diced

1 tbsp golden syrup

2 tbsp dark muscovado sugar

150ml double cream

Method

Preheat the oven to 180°C/350°F/gas mark 4. Butter two 20cm round sandwich tins, each 4.5cm deep, and line the bases with baking parchment.

Place the chocolate in a bowl and pour over the just-boiled water. Stir until melted, then set aside to cool.

In another bowl, sift together the flour, baking powder and bicarbonate of soda, then stir in the ground almonds. Cream together the butter and sugar until very light and fluffy. Put in the orange zest and mix well.

Add the vanilla extract to the beaten eggs. With the whisk running, very slowly add the egg mixture to the butter and sugar, adding 1 tbsp of the flour mixture as you go to prevent curdling, then add the cooled, melted chocolate and the buttermilk. Fold in the remaining flour mixture very gently and divide the mixture between the tins. Bake for 30–35 minutes or until firm to the touch. Leave for a minute or two in the tins before turning out onto a wire rack and leave until absolutely cold.

To make the icing, melt the chocolate and butter over gently simmering water, then stir in the syrup and sugar and gradually pour in the cream until all is well blended and smooth. Allow to cool completely, then whisk the mixture until it thickens. Spread half on the base of one cake. Sandwich the two cakes then spread the remaining icing on top. Decorate according to the theme of the celebration or the whim of the baker.

BEALTAINE

Living Hills

Casting Clouts

Early May has seen the return of dull, damp days and Sliabh Buí has been largely invisible behind a sheet of distant drizzle. Again this is apparently a good sign. It supposedly augurs a warm, fruitful summer – 'A wet and windy May fills the barns with corn and hay,' as the old proverb has it. I do hope so. Despite the chilly and cheerless weather, it has been impossible to wrestle the boys back into wearing vests – they have shed unnecessary underwear, and for them there is no going back. I told them of the old adage 'Ne'er cast a clout till May is out.' They liked the word 'clout' as a synonym for nether garments, and they particularly enjoyed the idea of casting them – they imagine people flinging pants about the place. I began to explain the ambiguity of the phrase and their eyes glazed over. Older Boy's favourite joke at the moment is one he got from his grandfather: 'Mum, what's a penguin?' 'Ask your father.' 'But I don't want to know that much about penguins.'

Still, I would like to be sure: should you cast clouts when the month of May is over, or do you wait until 'the' May is out – that is, when the May tree, usually the whitethorn, is in flower? Either way, we were a bit premature.

We are only a week into the month and our whitethorn has yet to show any sign of blooming; however, there is plenty of blossom about. The pear tree's flowers, like the plum's, didn't last very long, but the petals on the fruiting cherries are still clinging on, the apple trees are in bright-pink profusion and there are swathes of bluebells everywhere. They line the roadside verges of our hedgerow; there are clumps of them in the corners of surrounding fields; and along the floor of the oak forest, they float en masse in a haze, like a low-lying layer of cobalt-blue mist.

In woodlands, bluebells flower, flourish and dwindle during the brief period of light before the overhanging trees are in full leaf and the canopy thickens. We went for a walk to see them whilst they are at their best, which is now. The river has been well fed by rain water these past few months, so it was high and swiftly flowing. We stopped at a gently shelving shingle beach to skim a few stones before following the bank-side path past deeper, stiller waters. Long tresses of green and gold water weeds swayed slightly with the gentle flow of the peaty brown river. We turned in to the older part of the forest and wound uphill into the massive presence of dark, ancient oaks. Their buds are only beginning to emerge. High up in the light, the upper branches, tinged with a hint of green fur, are splayed in filigree against the sky. At ground level, the thick trunks loom out of a lagoon of bluebells, and as the clouds shift across the sky, patches of sunlight flicker over the flowers like ripples in lake water. Here and there between the oaks, young beech trees criss-cross like a trellis as they recede into the distance. Their slender trunks are covered with olive-green moss and their leaves are delicately coloured, but as the summer proceeds they will deepen into a darker shade of green. We walked further into the woods and waves of bluebells emerged from the shadows like large swatches of pale-purple carpet – I almost didn't notice the frochans in blossom amongst the undergrowth on either side of the path. Frochan is Irish for bilberry – or whortleberry, or huckleberry, depending on where you're from. In August, the small green bushes will be laden with small blue berries, but right now, hidden under dark leaves, their tiny pomegranate-pink flowers droop like miniature Chinese lanterns.

The path took us back down to the river past many fallen trees – victims of the February storms. Some straddled either bank and some lay around like the random angled poles you find in the ape houses in a zoo. One tree had been snapped clean off about ten feet from the ground, its jagged, splintered trunk on one side of the path, the remainder on the other next to a small, dark pond. The surface of the dank water was agitated so we investigated. It was full of wriggling tadpoles. A few years ago, this discovery would have merited at least twenty minutes of careful examination, but tadpoles aren't as fascinating as they once were – the boys looked, nodded and walked on. We strolled back along the river's edge. Warblers and buntings chirruped and squeaked. A duck and a drake – a mating pair of mallards – scooted about on the water, quacking happily. The warm sun shone briefly and Older Boy pointed out that if he and his brother had been forced to wear vests they would have been 'boiling'.

With the greening of the trees, the singing of the birds, the blossoming of the flowers and, most especially, the casting of the clouts, it is not surprising that the customs associated with May are the remnants of ancient fertility rites. When I was growing up in England, I first heard the term 'to go a-Maying'. I had a rather virginal idea of what this meant. I pictured girls in white frocks with flowers in their hair singing 'tra la la' as they skipped about in meadows picking wildflowers. I couldn't really see the fun in that, but it was a completely bowdlerised notion of what used to happen. May was a month of natural fecundity and human promiscuity. What 'going a-Maying' actually meant sounds much more interesting: young couples disappeared into the woods all night to 'bring the summer in'. God bless them. There is a poem by Kipling, part of which goes:

Oh, do not tell the Priest our plight,
Or he would call it a sin;
But – we have been out in the woods all night
A-conjuring Summer in!

In Ireland, the first of May, or the festival of Bealtaine, is – like all the others – associated with the sun. The major events of the solar year are the two solstices and the two equinoxes. The halfway points between these are the cross quarter days: Bealtaine, Lughnasa, Samhain and Imbolc. Bealtaine is midway between the spring equinox and the summer solstice, or Midsummer's Day. In Irish, the word means 'bright fire'. The day was sacred to Belanus – 'the Shining One'– a pan-Celtic god who brought heat and light. The coming of summer, however limp, to the land of winter, as the Romans named it, is a big deal.

Bealtaine is the first day of *samh* – the living half of the year – and is one of the ancient 'Fire' festivals. Great bonfires were lit both in celebration of the end of *gamh* – the dead half – and as a kind of sympathetic magic to rekindle and encourage the sun. Fire was believed to have transformative powers and represented transition from one state to another. On this day cattle would be moved to the summer pastures, ideally passing between bonfires on the way so that the smoke and ash purified and protected the herd. This was thought to have a restorative effect on humankind too – people ritually walked around the fire or leapt over the flames to take full advantage of all the available magic. Bealtaine and Samhain, or May Day and Halloween, were the two days when the gap between the human and fairy worlds was easiest to bridge. They were when the Sidhe were especially active and malevolent. This notion must have been suggested by the changing of the light as the sun waxed and waned. At Halloween the skies darken and the evenings draw in: the year dwindles and the sight grows dim. Threatening shadows and treacherous mists hinted at a subtle phantom presence. In early May birds rejoice and the blossom blazes: the returning sun brings bright mornings; the eye is dazzled back to life and the dead earth is miraculously revived. It is not hard to imagine sorcery at work.

Landscapes make their own mythology. The vast plains and tall skies of the American Midwest evoked Manitou, the 'Great Spirit'; the blizzards and ravines of the Himalayas conjured the 'Abominable Snowman' or Yeti; and here in Ireland, a single thorn tree can house a host of 'little people' – creatures whose diminutive stature is commensurate with the intimate terrain of our small island. The ancient Irish saw these agents of the supernatural everywhere and I almost believe in them myself.

I remember walking by a lake in Mayo in early summer some years ago. The sun was shining brightly after a recent shower and I saw a May tree standing alone between the path and the lapping waters. A few raindrops glinted on its sharp thorns and the light glittered on the still damp leaves. Gentle wisps of steam were rising in the warmth and, by a trick of refracted sunlight, scores of miniscule rainbows shimmered about the branches. It was beautiful, hypnotic and eerie. I could almost hear the tap-tapping of miniature hammers and the tinkle of tiny laughter. 'The little people' is a phrase that evokes images of tiny shoemakers and prankster pixies. It is twee, misleading and a touch patronising: ancient folklore dealt with something much darker. In Dublin in summer, Leprechaun outfits of green, felt hats and ginger beards are readily available at extortionate prices. The tourists snap them up and the traders make a mint, cheerfully conniving in the image of a charming peasantry simple enough to be spooked by mischievous midgets, but the original template of what was to become the twinkling leprechaun was much more spectral and dangerous.

The Sidhe were everywhere – in all weather and at all times – but were at their strongest from dusk on May Eve till noon on May Day, during which time there was powerful and potentially evil magic abroad. Lone travellers were waylaid by sudden fog and spirited away. The best protection was to wear your clothes inside out to confuse the phantom abductors. Babies were harder to disguise and were often snatched from their cradles. Changelings were left in their place and these would sicken, fade and die whilst the real infant thrived in Fairyland.

Sadly, the belief that the real child would be returned if the 'Fairy' changeling was destroyed sometimes resulted in infanticide. Whether or not this was a case of popular superstition masking the mercy killing of severely stunted or sickly babies who were anyway doomed to an early death is now hard to say. But the sinister myth of changelings hiding amongst us in mortal form still had currency almost into the last century and occasionally led to grisly crimes. In the year Marconi sent his first radio signal, a jury in Tipperary were told that the disturbed defendant had murdered not his wife, but a changeling who had taken her place. He was convicted and such beliefs faded, but the corresponding idea – that lost ones were happily cavorting with the fairies who had taken them – lasted longer. Perhaps the notion survived as a commonplace though meagre consolation to parents grieving for a dead baby – like the abundance of naughty cherubs in Renaissance religious painting. In times of high infant mortality, it might have been comforting to imagine a lost child fluttering mischievously around heaven or dancing by starlight with elves.

Aside from stealing children or infiltrating vulnerable families, most of the Sidhe's lesser malfeasance at Bealtaine was directed at dairy products. Petty trickery, perhaps, but to a population often on the margins of starvation, spoilt milk and bad butter were tangible losses. The defensive charms designed to thwart them were many and curious. Good milk would be poured on the thresholds of house and byre as a barrier against the supernatural. Spiteful fairies particularly enjoyed preventing butter from forming properly during churning and the best way to foil them was to chant a counteractive spell such as: 'Come butter come, Come butter come – Every lump as big as my bum.' My dad told me about this when I was small – I didn't believe him. I thought he must have made it up but he hadn't.

The safest thing to do on Bealtaine was to stay indoors and lie low. If it was absolutely necessary to venture out, protection was required. Carrying a piece of iron or a cinder from the fire was strong armour, but the best charm of all was to wash yourself in your own urine: the stronger the smell, the better the deterrent. Apparently it made the fairies recoil, and I am totally with them on this point. I would far rather wash myself in May dew, which was also common and a much more socially acceptable potion. This was a versatile and potent charm. If you collected the dew after dusk on May Eve and bottled it before the dawn you had an elixir with magical restorative powers for a year. If you couldn't be arsed to collect enough to bottle, benefit could still be got if you risked going out after dark: walking barefoot in the dew prevented corns and bunions; if a woman washed her face in it, she improved her complexion; and if she undressed and immersed herself in the dew-soaked grass between dusk and dawn, she beautified her body. A puritanically minded past president of ours liked to imagine Irish girls as 'comely maidens dancing at the crossroads'; I much prefer to think of them rolling naked in the fields on summer nights. If the idea of this alluring custom wasn't aphrodisiac enough, there were plenty of other rituals connected with fertility and propagation at this time of the year. Near the Bealtaine bonfire a pole was erected, and perched on top was the May Baby – a female figure draped with flowers and ribbons. A masked and costumed couple would dance explicitly beneath it and around the fire, gesturing lewdly – erotic pole-dancing is an ancient art, it seems. Childless women came from miles around to watch the hot action as an aid to conception.

This custom was a remnant of ancient tree worship – widespread in Neolithic societies across Europe. The English maypole is rooted in the same practice and I remember dancing around one during my first summer at junior school in Britain. My teachers had organised a May fair and we were all kitted out in white shirts and trousers and wore ribbon-festooned straw hats. We were given sticks covered with nailed-on beer-bottle tops, which jingled pleasingly when shaken. In a vaguely choreographed melee, we skipped and jumped and banged our sticks together. It was called morris dancing and it was very perplexing but quite satisfying. After this strange ritual, we each grabbed hold of different coloured long ribbons hanging from the maypole and hopped about, weaving in and out of each other as we went. When we had finished, the pole was covered from the top down in a symmetrical interlaced pattern of criss-crossed red, blue, yellow and green. I had no idea what it all meant, but the parents thought it was lovely. I now realise that myself and all the other nine- and ten-year-olds were dancing in formal adoration around a giant, symbolic phallus, and that our genteel, pretty ribbons had woven a colossal multicoloured condom.

The maypole never really caught on in Ireland. The fertility symbol more prevalent here is the May bush. Regional variations of tree were used but mostly it was the whitethorn, and in this part of the country that is what people mean when they talk of 'the May'. Bright ribbons, wildflowers and coloured eggshells saved from Easter were hung from the branches, and people danced in celebration of fertility, fecundity and, more recently, the Virgin Mary. In the eighteenth century a superior general of the Jesuit order in Rome grew tired of his students' profound immorality and declared the entire month of May to be sacred to Our Lady: he hoped this would encourage moderation, abstinence and chastity. Recalling my own student days, I doubt this was successful, but the association of Bealtaine and the Blessed Virgin spread, and by the nineteenth century most of Catholicism revered Mary as 'the Queen of the May'. Throughout the month, statues of her were crowned with flowers and paraded in and out of churches accompanied by children in stainless white, singing hymns in her honour. I have a pretty memory of such a procession from my own childhood, but even then the tradition was waning. Celebrating maidenhood in general, and Mary's maidenhead in particular, whilst the whole of nature is aroused and throbbing was never going to be a festival destined to last. We are too suggestible a species. How can we be expected to suppress powerful urges when everything around us is encouraging them? It is far easier, and much healthier, to surrender to the season and 'go a-Maying'.

Ten years ago on May Day, the Beloved and I were awaiting some friends from the city who were coming to visit for the long weekend. The spare room was made up, dinner was prepared and the wine was chilling. The friends rang to say they would be a little later than expected. The afternoon was balmy and we had time on our hands. I glanced idly at the paper while she picked a bunch of wildflowers for the kitchen table. Her hair was loose and she was wearing a light cotton dress which hinted at the contours of her body. She noticed me looking at her and her eyes glinted. My mouth went dry. She went upstairs and I followed. That was Bealtaine. Older Boy was born the following Imbolc. In subsequent years, we have noticed that he has quite a few friends with birthdays near his own. It seems we were not the only people invigorating ancient traditions that particular May Day.

Traditions evolve and change as new cultural influences make themselves felt – a process observable as we have become more and more integrated with Europe and accelerated during the Celtic Tiger years. I noticed it happening when I first came back in the dying years of the last century. I had landed my first job at the national theatre and was thrilled with myself. Early in rehearsals, I left the building during the midday break and strode forth to see what Dublin offered in the way of snacks. I saw a blackboard advertising a lunchtime special of a rasher baguette and a pint. The wind of change was starting to blow. It later grew to hurricane strength and has since subsided, but the culinary cross-fertilisation is now nearly total. Nettle soup was once a seasonal dish in May (I have tried it and wasn't hugely impressed) but the other day Female Farmer Friend called in for coffee and brought a jar of her freshly made nettle pesto, which was delicious. We had some with pasta for lunch the next day.

She also brought sad tidings – Philomena is dead. I was sorry to hear it but not surprised. I have wandered into the barn once or twice over the last few weeks when taking the boys for walks up the hill, and I had seen for myself that the creature was languishing. The medication failed to cure her blindness, the teeth grinding got worse and the poor thing stopped taking milk from her mother. The vet had done the merciful thing a few days previously. Female Farmer Friend was hesitant to break the news for fear of upsetting me – evidently, in her eyes I have yet to acquire the grim stoicism of the true cowman. Her husband seems to think I am made of sterner stuff because he often sends me strange and gory pictures which he thinks might interest me.

Last week we had some friends from the city for lunch: a flamboyant gay tenor and a more flamboyant soap actress. The Beloved and Soap Star were cackling at Gay Tenor's stories of tricky divas he had known when my phone buzzed. It was a graphic photograph of a heifer undergoing an emergency caesarean section in the cowshed. I resisted the temptation to show the company the shot. It might have dampened the mood somewhat and I would have found it difficult to explain. I can only speculate on the consequent tales that might have grown in the telling. Many of my creative friends and colleagues are adroit at spinning minor incidents into elaborate stories. It is a habit of mind of which I thoroughly approve: I do it myself – I am doing it now. On my subsequent visits to the city, I might have encountered rich rumours about my bizarre taste in pornography.

Still … it might have been amusing to pass the picture around whilst nibbling prosciutto and pâté. Having gutted various creatures prior to cooking and eating them, I have become inured to the sight of animal entrails. But the main reason that the sight of a cow's opened abdomen doesn't particularly bother me is that I have had my own abdomen surgically sliced, more than once, so a little bit of bovine blood is no big deal. I was ambushed by severe and chronic illness a dozen or more years ago. My health is now generally tolerable but there are occasional dips, and this week, I finally got a low-level but persistent problem dealt with.

Before I got into the car to head up to the city for the required minor operation, I watched the swallows zip in and out of the narrow gap between the doors of the garage where they are nesting. The speed and finesse of their flying is astonishing. They hurtle towards the inch and a half wide crack, almost too quick to see, and at the last second they turn sideways so that their outstretched wings are vertical and they disappear into the darkness within. Their eyesight instantly adjusts to the gloom and they land abruptly and precisely on the nest in the rafters a few feet inside the door. Sometimes I surprise them going into the garage (it's where Mellors keeps his tools) and they dart out of the shadows and over my head to the wide expanse of light and air outside. It is wonderful to

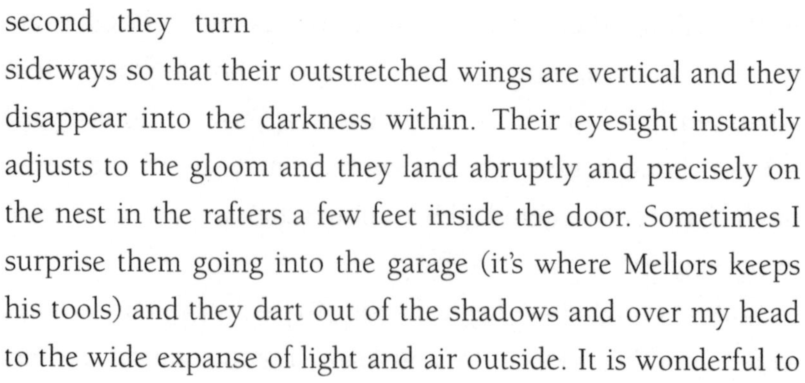

have them year in and year out, except that they crap all over the handles of my spade, hoe and rake. There is a price for everything.

The lilac bushes by the vegetable garden were beginning to flower, the apple trees were in pale-pink blossom and in the hedgerows the blowsy blooms of cow parsley were nodding their big white heads. I started the engine and drove off cheerfully to submit to the knife. My mood was still buoyant when I arrived at the hospital – in fact, it even lifted slightly as I checked into the Day Care Centre. I know that this sounds odd. Most people, quite reasonably, think hospitals depressing places – they are, after all, full of the sick and dying – but I find the sights, sounds and smells of this particular hospital reassuring. I was so well looked after here when I was really sick that I feel safe and comfortable when I return. This mood seemed to be general when I came round from the anaesthetic a few hours later – the other patients in the Minor Op. Recovery Room were exchanging good-humoured, ribald remarks and I joined in as soon as the lingering grogginess had passed. Everyone's procedure had gone smoothly and I suppose we were all relieved we hadn't had to undergo anything more serious.

When the nurses brought tea and toast for all, the atmosphere verged on the festive. I have had many a tea and toast in this hospital before and I have always been grateful for it, if a little bemused. For one thing, the teapots puzzle me. I am in a building which is full of complex, highly advanced lifesaving diagnostic technology – machines which demonstrate extraordinary levels of human ingenuity – and yet we seem incapable of designing a mass-produced teapot that pours properly. Then there is the steamed toast. I open my little pat of butter and my little pot of jam, salivating slightly as I eye what looks and smells like freshly toasted slices of brown bread – I pick one up and it wilts in my hand like a punctured lilo. It is warm and dry but also soggy. I don't know how they do it.

Having had a general anaesthetic, it was not recommended that I drive, so when I was discharged I was met by a fond auntie and spent the night at her place before heading home the following day. Since we lost our city centre flat, I stay with her when I am working in town and she always makes me feel most welcome. She is well plugged into the cultural and political life of the city. She is familiar with the theatrical milieu, as her son, my cousin, was an actor for some years before he saw sense and got a proper job. He is now a highly qualified counsellor and therapist, and he took his first tentative steps in this direction by giving people a psychological profile based on their personal horoscopes – an occupation with probably more job security than acting. Fond Aunt is also a veteran of many social and ecological campaigns over the years, so we have plenty to chat and gossip about and I always enjoy staying with her, which is good, as we will be seeing a lot more of each other over the summer. I have been offered a job at the Abbey and I am delighted.

The job came out of the blue – a joyful surprise. Usually days and sometimes weeks can pass before you hear anything after an audition – this allows plenty of agonising time to rerun the meeting in your head, reinterpreting every nuance to deduce that you definitely got the part or that you definitely didn't. I do this, lurching from elation to anguish, until the Beloved remembers that she really has to clean the grouting in the bathroom. This was an unexpected gift of an interesting role in a lovely play by an eminent living writer, directed by someone I know and like and at the national theatre. There was nothing to do but jubilantly yell, 'Yes!' down the phone to my agent and then bask in what is always the best bit of any job – the period between getting the part and actually beginning the work. This will be a short enough time in this case because rehearsals begin before the month of May is over. I thought this play had already been fully cast – perhaps it was and someone pulled out. I don't care – I will be working again soon. Hooray!

In the meantime, there is much work to be done before I leave for the city. I must mow the acre. I must propagate sweet corn so the plants will be ready to plant out in June. I must plant beans and more salad leaves and carrots. Each year I make a diagram of the four beds in the vegetable garden and map out what is to be planted where. This changes annually – some sort of crop rotation helps the soil to rest and prevents particular pests getting too established. Making the chart began as an aide-memoire and has now become a ritual. When I sow, I make an appropriate addition to the diagram in pencil on the blank paper – as though I am thereby calling the plants into being. And then, as March, April, May and June pass, I can see the progress of my labour as the rectangles in the drawing are gradually filled in and the plants in the beds start to poke up from the earth. Then, best of all, in July and August I see my plans made manifest as the season of abundance returns ... But that is months away yet, and right now the pressure is on. I need to hurry. Mellors must make his map. Lesley the lumberjack also has urgent work to do. The stove is still being lit once or twice a week. If the sun is shining, May can be warm and glorious, but on a wet or sunless day there is a lingering chill. The boys may have cast their clouts, and if it gets too cold I suppose Perkins can always wash them again, but we still require fuel to burn so I have to make sure there are enough logs chopped should the Beloved need them. Besides, with summer coming, it will soon be time to get out the barbecue and we need to have the wood ready for that.

NETTLE PESTO

Ingredients

150g young nettle leaves

120ml olive oil

30g pine nuts or walnuts

2 cloves garlic, peeled and crushed

55g freshly grated Parmesan or other very hard cheese

salt and pepper

Method

While the whole pestle and mortar carry-on is hugely reminiscent of Mama in her kitchen, I am all for modern conveniences. So put the nettle leaves, olive oil, pine nuts and garlic in a food processor and pulse until you get the consistency you like. Move to a bowl and fold in the grated cheese. Taste and season. You might not need any salt depending on the cheese you use.

Keeps well in the fridge for up to a week – if it lasts that long!

Learning Lines

There are many songs in the Irish repertoire that are set in the merry month of May, and most of them seem to concern a journey. Many have the same opening line: 'As I roved out one bright May morning'. My head was full of them as I hit 'the rocky road to Dublin' to begin rehearsals. It was a 'bright May morning early' and my heart was 'a-bubbling'. The field was freshly mowed; the kitchen garden was as it should be for mid-May; and there was enough wood chopped for both the stove and the barbecue. I think I can safely say that I have achieved the totemic task I set myself. We have got through the year, summer to summer, without paying a cent for fuel for the stove. All it has cost us has been fuel for the chainsaw … and my labour, of course, but that has been gladly spent. In fact, that is the essence of my satisfaction. My effort, sweat and occasional cursing have benefited my family in a clear and tangible form – I have kept us warm through the winter. My professional life has been dormant for most of this past year and sometimes it is hard to keep a sense of self-worth when you are middle aged and out of work – but I kept the fire lit. A small thing, but I am pleased. However, I am not sure my inner lumberjack has ever been fittingly named. The Beloved's notion of 'Lesley' as a slightly embarrassing, and therefore rarely mentioned, name amused me and I adopted it, but it never seemed to sit quite as well as Mellors the gardener, or Quigley the handyman, or even Perkins the laundry drudge. It felt somehow wrong because the right name is, perhaps, my own. Lesley – *c'est moi*.

Before I set off for the city, I took a stroll around the acre. The blooms on the pear trees had dwindled but the apple blossom was thickening. In the vegetable garden the potato plants were looking sturdy, the broad beans were stretching up and the climbing peas snaked around the tilted canes. Across the valley the recumbent mass of Sliabh Buí squatted on the horizon. I see the yellow mountain every day, often without noticing it, but sometimes it gently grasps my total attention and halts me. The details of its contours were sharp and vivid in the intensely bright sunshine: the gradations of green and blue, with cleanly defined edges, looked like a giant natural collage. Not for the first time I felt I was the still point in the centre of a living picture. I stood for a moment, absorbing the unexpected instant through all my senses even though it was primarily experienced by sight … and the radiant light.

The Latin word *claritas* popped into my head. I have come across it in the work of Heaney and Joyce and, directly translated, it means clarity, unsurprisingly. But it also has an aesthetic connotation – a kind of effulgent splendour which confers delight

to the intellect via the senses. It derives from the definition of beauty suggested by the medieval monk Thomas Aquinas. He thought this sense of *claritas* was a necessary component of anything considered beautiful – *id quod visum placet* or 'that which pleases on being seen'. Then I thought, can this beauty around me still exist when it is not being seen? When it is not being actually looked at? Is beauty only ever there as merely a perpetual potential, waiting for the right light and the right viewer? As I stood looking from that particular place at that precise time – a never to be repeated, unique, subjective moment – I wondered if beauty can ever really be shared. Then I realised that my profound and important insight could be expressed as 'Beauty is in the eye of the beholder' … So, nothing new there – my pondering could be encapsulated in a cliché. I got into the car and turned the key in the ignition. The spark plug fired in the same unremarkable fashion as the synapses in my brain. I pulled out on to the road and glanced back to look at the lilac bushes in full flower behind the gate – tall, expansive and vividly, well, lilac.

The high green hills parted in front of me to take the road. Hardly a cloud in the clear sky, and those few were high. The woods and grasslands rolled away to the horizon. I inhaled the open view in slow, steady, expansive breaths and smiled as I drove. In the hedgerows that lined the road and partitioned the fields, the whitethorn was beginning to blossom. I wondered if the May would be out by the time I was driving home next weekend. I idly contemplated thorn trees as I followed the familiar route. When they are in leaf, I find it hard to tell blackthorn and whitethorn apart, but they flower separately and differently. The blackthorn is first and its blossom opens before its leaves do. The pale flowers appear in clusters along the bare black branches, like blobs of cotton wool on a splayed bunch of sharp black sticks. It has been profuse this year, so hopefully there will be plenty of its fruit, the sloe berry, to steep in gin this autumn.

When the whitethorn flowers, the tree is already in leaf; the buds of curled-up petals nestle among the leaves like white peas in a bed of basil. When the blossom bursts into balls of flower very like the blackthorn, but bolstered by the greenery, it looks more substantial – the trees are like giant white bouquets studded about the landscape. Or they will be soon enough. Traffic thickened as I approached the motorway. This was exciting rather than annoying. I examined my optimistic mood for traces of guilt at abandoning my family. Finding none, I wondered if I should feel guilty about my apparent selfishness. 'Nah!' I thought, as I eased onto the slip-road, geared up and sped towards town.

From the theatre's high, bright rehearsal room on the top floor of the building, the city looks good in the early summer light. Above the noise of traffic and pneumatic drills, you can see across the rooftops where pigeons roost to domes and spires and towers. Between gaps in the grey buildings there is a glimpse of the river glittering as it winds inland. I am staying with Fond Aunt and her house is a short tram ride from the city centre. As I make my way back to the stop each evening, every pub, cafe and restaurant I pass has tables, chairs and awnings outside; and people eat, drink and jabber in the sunshine. This al fresco socialising is a recent addition to the city's life, as is the tram system. I was watching Fellini's great film *Roma* some years ago. There is a wonderful sequence where workers on the Roman underground railway are extending a tunnel when they inadvertently uncover a buried chamber, unseen since ancient empire days. Extraordinary frescoes are revealed by the workmen's torches. Just as they are seen by human eyes for the first time in centuries, the newly admitted air, in what had been a sealed environment, changes the atmosphere and the gorgeous images wither and fade. It is an event that actually happened, and Fellini imagines and renders the tragic moment with a fleeting beauty. In the ensuing dialogue, one of the engineers talks about his next gig and his plans to head off to work on a planned new light-railway system in either Munich or Dublin. The film was made in 1955 and, with typical Irish urgency, we got the Luas, as it is called, a mere fifty years later. Perhaps the city fathers were not keen on admitting to a mistake, as the original and extensive tram system had only been closed down a decade or so earlier. Anyway, we now have two spanking new tramlines, one on either side of the river, and I enjoy using them. At lunch time during this past week, I have taken the lift to street level, hopped on a tram for a couple of stops, done a little shopping in the Asia Market, had a bowl of spicy fried rice and then nipped back to the theatre. These are palpable delights I was once blasé about. I am again an ingénue, up from the country and marvelling at the amenities of the city.

The first week's work has consisted of reading, discussing and exploring the play. I love this initial process, before the harder tasks of learning, repeating and performing start, and I am lucky to be able to do it. When the work is good, it amounts to sitting around in a relaxed but engaged environment with a bunch of clever, creative people and talking about a great work of art. The director is illustrious: he has enjoyed many great successes and he doesn't need to prove himself. He approaches the play with respect and without the desire to 'interpret' it in any way. He wishes merely to elucidate it and he has assembled a great cast to do the job. I am secretly thrilled to be amongst their ranks but I affect nonchalance. I know most of them and have worked with many. This is going to be a thoroughly enjoyable gig, with all due seriousness of mind but also plenty of laughs. On one side of the table is a young actor playing one of the central roles – he has a big talent but a small ego and doesn't seem too intimidated by the fact that across the table from him are two venerable and hugely respected actors, both of whom have played his part in the past and one of whom, in fact, created the role when the play was premiered here thirty-five years ago. This is one of the things I adore about this theatre – there is a sense of tradition and history that feels supportive but not stifling. The torch is always being passed on to succeeding generations. These two elder statesmen – one venerably bearded and the other venerably balding – watch the young pup with a benign, appreciative eye. They are not interested in bringing up their performances in the role because they are more concerned with the task in hand – which is to create the characters they are playing now.

Watching other actors discover and develop their parts is fascinating and often useful. You never know when you might observe something you can steal. Every actor has their own special qualities and idiosyncrasies but ultimately we are all engaged in the same task: to imagine and then present a plausible living person who does and says whatever the script requires them to do and say. In a superbly crafted play like this, the text is where you start and end. You explore what is said, unsaid, overheard or implied. You mine these layers of meaning to explore the emotional texture of the piece. This is best done collectively. Being open to suggestion from others offers you lots of material to nick. This is called 'being influenced by' someone. Then each actor, individually, and in a kind of psychological crossword puzzle, builds up an interior picture of their character. You try to create an alternative conscious and subconscious mind, which you then attempt to inhabit during performance. Doing this requires empathy – after all, you are trying to explore what it feels like to be another human being – and also self-examination. It is often easiest to start with the personal and use details of your own experience to realise your role. So, at heart, what we do is use elements of our private lives to create an illusion which, hopefully, in front of an audience, presents a shared perception of an imagined truth. We use lived life to make an artifice which reflects reality. In many ways it is an absurd job for a grown-up. But I love it.

This approach only works when the play deserves it, and this one does. I talked earlier of performing in a play – twenty years ago in London – set in rural Ireland during the month of August. The play I am investigating now is by the same writer and is set at the same time of year and in the same fictional village in remote Donegal. It is played out against the background of a crumbling mansion in which a once powerful family of Catholic gentry have been dwindling for generations. A wedding is imminent but a funeral occurs instead and a moving family tale unfolds. But, as in all great plays, the story is a metaphor which works on more than one level, and behind this private drama a larger public play looms. The whole work is a subtle and profound meditation on the narrow-minded and ultimately repressive power structure composed of unquestioned religion and mythologised nationalism which threw its long shadow over the country for decades. Well, that's what we reckon anyway … I wonder if the critics will notice.

When I finished the first week's rehearsals on Friday evening, I was tempted to join the rest of the cast in an al fresco pint but I was eager to get home. The city was hot and clammy and I crossed the river at a brisk pace in order to escape to a gentler one. It was a lovely early summer's evening and the drive was perfect. A clean yellow light made the motorway cheerful, and when I got onto the country roads, the fat hills around me looked tangible and succulent. The humps and hollows were draped in folds of blue and green baize. There is no high summer haze yet so I could clearly see for miles – distance and detail simultaneously. The May was indeed out. The full-blossomed trees looked like huge meringues piled with whipped cream and the fields appeared to be lined with giant sugar sculptures. I love where we live. And so does the Beloved. The current high-toned beauty of our environment must compensate her for the extra tasks that fall to her whilst I am away. Or so I reasoned.

As I drove up the hill I saw that the ploughed fields on the slopes of Sliabh Buí were now fully green but there was still a touch of yellow from lingering gorse. I got out of the car and birdsong pierced the silence. This was quickly drowned out by a clamorous welcome from the boys. They hugged and kissed and yelled that I was just in time to take them to football. I told them I had hurried home for no other purpose and got back into the car. Down at the Fair Green the adults watched the children play. The other parents seemed oblivious to the mighty and tumultuous work I had been engaged in during the week, so I thought myself back into being Mr Vice-Chairman. Over the weekend I felt I should make up for my absence and I revisited Perkins and Mellors. They are not as amusing to play as they once were, but nonetheless I did much laundry, mowed the acre, transplanted the sweet corn and sowed turnips, carrots and purple beans. Perhaps I am feeling guiltier than I thought. I did most of these chores on Saturday in time to take Older Boy to his golf lesson on the course which now occupies what used to be an old aristocratic estate on the other side of the village. While the boy was swinging his clubs, his younger brother and I had a look at the house. It is a big Georgian pile of buttery sandstone and is exactly the sort of place we are evoking in the play. Right now it is surrounded by an explosion of circus colours. A profusion of mauve, red, purple, puce and yellow rhododendrons line the walls and dwarf the passers-by.

Sunday was the grand opening of the village playground. The Beloved has been an active and effective member of the committee that raised the money, got the planning permission and hired the builders. It is very nicely designed, constructed out of wood and ideally positioned on the village green which, though central, has not been much used recently – the playground might give it focus. The money was mostly raised last summer, it was built during the autumn and it was finished as winter set in. There has been a wait for finer weather to open it and also in order to let the grass grow up around the various swings and roundabouts. (There is also an abseiling zip-wire kind of thing, which I am very keen to try.) The weather has been lovely all week, but clouds had been gathering during the morning, and as soon as the parents and excited children assembled, it inevitably poured down. The kids didn't care a bit – they charged in when the ribbon was cut by four children chosen at random. Then they horsed around delightedly, getting thoroughly drenched whilst the adults loitered and shivered under umbrellas.

The last weekend in May followed a similar frenetic pattern of mowing, planting, laundry and sport. On the Sunday we had a festive family lunch to celebrate the Beloved's mother's birthday. The Beloved's sister was also in attendance. Unable to drive because of a recently broken leg, she had come down in the car with me for the weekend. Her idea of a good time is to exercise top-flight racehorses at top speeds. She was flung from a highly strung galloping filly but is quite sanguine about her injury – she considers herself lucky she didn't break her back or crack her skull. Her nephews think she is cool and her mother thinks she is mad.

The birthday girl was delighted to have both her daughters and both her grandsons with her to mark the occasion. She adored the homemade gifts from the boys: the younger one gave her a hand-painted clay snake on a yellow plate (to represent sand); and his brother gave her a butterfly mask made from a cereal box, which he had covered and coloured. We had a cheerful Côtes du Rhône with the cheerful meal, and with the cake we drank a bottle of dessert wine from Bordeaux. When we played Stevie Wonder's 'Happy Birthday' loudly in the kitchen, all the adults started throwing shapes in their chairs. There was soon a seated disco scene, which the boys found excruciating. At one point Younger Boy had his head in his hands as he murmured, 'If you are ever going to shoot me, now would be a good time.' His older brother just pleaded with us to stop.

The city has been baking in the early June weather and the sun has been splitting the pavements. All the outdoor tables of the bars and cafes are full of people dressed for the beach. Everywhere you look there is vivacity and colour – the dominant shade being the glowing pink of traditionally pale Irish skin. I have been catching up with a few city pals, but mostly indoors: one evening I brought steaks and frites over to a friend's apartment and he dusted down some superb wine from his excellent cellar; another night was spent over creamy pints of stout in the dark, cool interior of an old Dublin pub. On one occasion I hopped on a bus to a friend's house for supper. As we chugged along the North Circular Road – a wide street lined with substantial Victorian houses – people were lounging on their doorsteps in T-shirts, chatting to their neighbours. This friend is the mother of a small boy younger than my older and older than my younger. We ate early so he could join us and it was still light when I was leaving so I decided to walk back into town. I turned off the main road and down a quiet avenue of solid Edwardian red brick. Trees on either side were in full leaf and threw a dappled shade as I strolled. The avenue ended and the masonry changed. I passed an old nineteenth-century asylum – a gaunt grey edifice topped by a cupola and clock tower. Long closed, it has since been used as a location for various movies and is currently being renovated to form the main campus of one of the city's universities. What a great place it will be to be a student. On a slight prominence and with open grounds, it has an expansive view across the city.

As I continued down towards the river, again and again I saw the modern city overlaying the old – an architectural palimpsest. Twenty-first-century flats, with washing hanging on balconies and the grime of traffic on their terracotta colours, loomed over Edwardian houses and Victorian cottages. The concrete curves and sharp angles of the recent constructions clashed or chimed – depending on personal taste – with the faded brick and ruled lines of the earlier buildings. I walked on down into a large, open square. I lived not far from here when I first came back to Dublin.

There used to be a market where you could buy huge sacks of vegetables for next to nothing – a sack of spuds for the price of a packet of crisps. I often made this false economy – the vegetables would rot before you could get through them all. I remembered the horse market that used to be held here once a month. You wouldn't exactly see thoroughbreds on display but there were all manner of ponies, workhorses and nags changing hands with people, buyer and seller, actually spitting on their palms to seal the deal. I saw it happen. Now the square has an arthouse cinema, delis, restaurants and a playground skate park set amongst the cobbles. Not all has changed, though – I spotted a few untouched pubs still trading briskly. I bought a choc ice, crossed the tramlines and emerged at the river by the grand old Georgian court building which was shelled in the Civil War. I leaned on one of the warm stone balustrades that line the banks. Gulls squawked and I caught a whiff of the sea from the tidal mud caked at the water's edge. The lowering sun glistened and dazzled in gold and silver ripples on the river's oily swell. My gaze sought the cool, soothing contrast of the deep black velvet shadows underneath the arches of an old granite bridge. Everything looked fresh and new, but also familiar.

I had seen all this before but why hadn't I quite noticed it when I lived here? Why did I not pay such attention to the particular? Perhaps I did and have just forgotten. Or maybe it is because living in a city puts scales on your eyes. In my troubled teenage years in an industrial English city, I kept my eye inward and my head down. London was thrilling but sometimes threatening so I kept my guard up. I also viewed the capital through the ideological prism of my politicised youth – I saw the architecture of power housing the booty of an empire. Dublin felt more convivial on my return but I explored it in a conscious process of reappropriation.

I was reclaiming my patrimony: the land of my fathers – and mothers. Only now, having made my own home, having staked and upheld my claim to an acre of Wicklow, can I look around me with relaxed, unqualified interest and view elsewhere with less tendentious eyes. A huge gull dive-bombed me. I leapt and dropped the remains of my ice-cream. The gull swooped and caught it before it hit the water. I swore and then took a moment or two to regain my composure. You really have to be alert and

streetwise in the city. When my breath was once more even, I crossed the bridge. I looked upstream at the huge Guinness brewery, which has been part of the city's fabric for centuries, and downstream at the new bridges, which have spanned the river for a few short years. Then I wandered through the 'cultural' quarter much maligned by many Dubliners, but vibrantly cosmopolitan to born-again hicks like me. The quality of buskers has certainly improved in recent years. I listened to a few before catching the tram back to Fond Aunt's house to learn my lines.

Rehearsals are going well and the director seems pleased enough with what I am doing. This, in turn, pleases me: it is gratifying to gain approval from peers, especially directors – after all, they are the ones who offer you work. I love doing what I do and I like being in relaxed control of my own performance, but other people have to enable this. I need the director to employ me and then give me permission to take charge of what I do on-stage. I have no doubt this need for affirmation from authority indicates a residual subconscious urge to seek validation from a father figure and thereby assuage my inner child ... blah, blah, blah ... it is all too complex and dull to investigate, but there is one reason that I especially love the theatre – it is an environment which encourages and usually requires the expression of powerful feelings.

I spent most of my teenage years suppressing them. A deep unhappiness accumulated and periodically burst out in unpredictable ways. There was the time when I turned twelve and got a new football strip in the colours of one of the city's two teams. I didn't really like soccer and wasn't very good at it, but I tried to join in. And besides, supporting one of the teams theoretically halved the number of bigger boys who might want to beat me up. I was a patrol leader in the scouts at the time. On the day of my birthday our troop was playing a match in a local league. I wore my new strip and played badly for the first half. Unsurprisingly, I was substituted at half-time and I burst into tears. My cries of 'But it's my birthday!' and 'I've got a new strip!' were to no avail – I didn't go back on. At the next weekly scout meeting, there was a strange buzz in the air when we arrived. Something was up. There was a long trestle table set up against the wall where we usually saluted the queen and made the scout promise. The meeting was brought to order and I was called to the front. A court was in session: for the recent display of tears, which indicated a lack of leadership qualities, I was to be stripped of my patrol leader's stripes and reduced to the ranks. I managed to clamp my quivering lips shut and avoid tears whilst I endured this humiliation, then I stepped back to rejoin the silent troop.

For the next however many weeks – I can't remember – I applied myself diligently but joylessly, and in due course I was re-promoted. I went home that night and never went back. I kept the incident to myself but I felt a deep sense of injustice at this use of public shaming to admonish a display of feeling. No doubt this memory has something to do with wanting to spend my life in an occupation which requires and applauds such a thing. Perhaps my entire career is based on a desire to give the finger to a middle-aged scout master, but looking back, I also have to acknowledge the possibility that my whinging at the football match was really, really annoying.

Over-indulgence in emotion is infuriating to witness and should be avoided. As a young actor I was eager to be overwhelmed by passion in performance. Then I grew up and copped on. The writer of the current play scrupulously avoids the same hazard

– to such an extent that he has decided to change the ending because apparently he has felt, for some years now, that the original conclusion hinted at sentimentality. This is quite thrilling for the cast – and, indeed, the whole building. The Abbey is a theatre built on the primacy of the written word: the founders, W. B. Yeats and Lady Gregory, were both writers, and here the author's vision is paramount. This is exemplified by the reverence paid to the texts of this particular writer. To hear that he is going to make a change is a source of great excitement to us all. It is merely a shift in emphasis, not a full rewriting of the scene, but nonetheless it will be new, and we will be the ones to do it for the first time. Well, I won't be – my character leaves before the final moments of the play – but I don't care: I am still involved.

I watched the rehearsals of this adapted finale the other day. It is now stronger, more honest and consequently very moving. The play still ends, as it did, with some of the characters humming and then singing a verse of an old song which has buried resonances for all of them. It was a beautiful resolution, but with judicious changes, a mournful, lyrical and almost lachrymose tone has been replaced by a much more strident mood. A piano swells triumphantly and the tentative voices grow in strength as the bruised, damaged people decide, just for an instant – just for now – to keep buggering on. They defiantly determine to be almost happy. I watched the moment being created and I wept like a baby: partly because it was great and partly because I was not in it. Maybe I haven't grown up that much. On one level, actors never really grow up. What we do is called 'playing', after all, and we are often easy targets for bilious hacks who like to dismiss us as emotionally incontinent 'luvvies'. It is true that actors are generally poor at talking of what they do without gabbling pretentiously about theatre saving the world. The thing is, it just might. I play an American academic researching the history of the family's forebears, whilst watching the current generation as they stagger through life. I see their failings and discover their losses. Watching the final moment of the play being rehearsed, observing the characters sing together as they briefly glimpse a unity of purpose and accept and support each other, I thought of how empathy is possibly the finest human attribute. Seeing the world through someone else's eyes, imagining and understanding their pain – being moved by it – can maybe make one pause before dismissing strangers' lives as irrelevant. Most of us seem to spend our time largely indifferent to others and separate from the massed ranks of humanity. Foolish as it may sound, I really believe that art can bridge the gap rather than exploit it … once in a while. This is why I stubbornly persist in trying to make a living out of acting – that and the dressing up.

Hunting Crabs

I was let off early one weekend so that I could get down in time to see the boys' end of year school concert and, despite loving my job, I still felt the thrill of the truant as I hit the road. The drive down was again lovely – plump green hills surmounted by plumper pillows of white clouds. When I pulled up at the school, the surrounding trees were in full June leaf, waxy and glistening; and after the background hum of the city, I was sharply aware of the clarity of birdsong from above and around in all directions. The Beloved emerged from the side door of the school – she has been in charge of the choreography of a rap number – and we took our seats just in time for the show. It was epic.

The story told of two tribes divided by an angry river. One tribe was tough: dressed in black, they had their faces painted with angular, geometric patterns. The other tribe was arty-farty: they wore multicoloured flowing robes and had floral-patterned make-up. Each tribe declared the other tribe to be rubbish. Then a TV crew pitched up. Led by a PR dolly dressed like a pink meringue, they decided to interview the tribes and build a bridge over the river. As you do. The bridge merely allowed inter-tribal insults to become more up close and personal. War threatened. Then the king of the arty-farties had a baby (or perhaps his wife did) and he offered it to the tough tribe, whose hearts were melted and all ended peacefully and joyously. Our boys were in the tough tribe and both played their parts with gusto. The younger one was fierce and proud and looked delighted to be up there. He kept grinning out front, even when his attention should have been on the action – quite a change from the vexed embarrassment of his Christmas-show appearance.

There were interludes along the way. There was a dream sequence represented in silhouette on a backlit bedsheet, with hand-held torches and puppet shadows. There was a chorus line made up of the TV crew and the bridge builders in yellow hard hats – the high kicks were erratic but they all had very good jazz hands. There was the de rigueur massed penny-whistle number – this time it was 'I've Been Working on the Railroad'. But my personal favourite was the rap. The Beloved's choreography was slick and edgy and the 'crew' were tightly rehearsed. They were a troop of hip-hop bears, who turned up for no obvious reason, slouched around to the beat of bass and bodhrán, rapped their lines smartly and then slouched off again, never to return. The head of the posse, or 'Mr Bear', was Older Boy and he was cool. He snapped out his lines with attitude, and his twists and steps were precise and sure. He also looked sharp. I noticed he was wearing my black trilby, my black leather blazer and my good black tie. His mother had supplied his crisp white shirt. The finale was an inter-tribal rave with much chanting, stamping, twirling and ululating. Afterwards, when I was proudly congratulating them both, Younger Boy asked me, 'Are we good at acting?' 'Yes!' I replied. 'You're great!' He went on: 'As good as you?' I hesitated – I had two conflicting impulses: 'Nearly, not quite, but you will be.' Older Boy pointed out that their show had been practised and presented in four weeks, and mine was going to take six weeks' rehearsal by his reckoning. He said no more but his implication was clear – we professionals up in the city were obviously slacking.

Over the weekend there was the usual rush of work to be done in my brief break. The sweet corn I had transplanted into larger pots a week or two ago and had since been brought on in a friend's polytunnel now needed to be planted out in their final growing positions. I also needed to germinate more – Younger Boy is tricky with vegetables but will eat corn on the cob with relish, so I grow as much as I can. I also planted cucumber and courgettes and rigged the canes for the beans to climb. As for planting the various varieties of beans themselves, I am going to have to delegate that task to the Beloved. The pressure to plant feels inappropriately intense – time is short and it usually isn't. I feel a hint of anxiety about my future absence. Will the Beloved step up to the mark with the vegetable husbandry? I must be brisk and brusque – there is now an inevitable urban influence on my sense of time management. I feel rushed and therefore slightly out of sync with the country – just as I often feel out of step in the rattle and throb of the city.

Cutting the grass in the acre helped me to gear down a little. Trundling up and down on the borrowed sit-on mower passes a pleasant, meditative hour. Scything through the lush green growth, backwards and forwards up and down the field, is repetitive and relaxing. With a warm sun on your back and a cool thigh-tickling breeze up the leg of your shorts, it is delicious. Mowing the orchard requires more conscious effort – you have to be creative. I cut a swerving swathe around the trees, reversing, ducking under branches or lifting and bending them to manoeuvre past. The cherry trees are thick with fruit – already, some are beginning to wither like they did last year, but a larger proportion seem to be swelling. There will be plums by the bucketful, whether they ripen or not. There is one pear forming, but this time on the other tree. And again there will be apples galore.

Late on Sunday afternoon I split a few ash logs into small chips and lit the first barbecue of the year. For the first time, the boys were keen to help – they twisted the newspaper into coils and piled up the twigs and wood chips and I let them strike a match each to get it going. It smoked profusely for a bit and then it caught. As I waited for the flames to die down, and for the wood to turn to a slow-burning pile of uniformly grey ash ready for cooking, I strolled around the freshly cut grass with a cold beer. The swallows dived and swooped then buzzed through the warm air, fat with flies, a few feet from the ground. I wandered through the fruit trees, picking up any small branches I had broken off during mowing before the Beloved noticed. I plucked a dead twig from one of the trees as I passed and got a sharp and painful wasp sting – my first of the season. The fecker was nestling within the blackened leaves. Every special moment has a tax.

Polytunnel Pal came over to join us, bringing more of my hardened-off sweet corn and a couple of tomato plants she had grown from seed. I planted them and then we ate. Slow-cooked homemade burgers and some local sausages, all of which acquired a delicious smoky flavour from the smouldering wood chips. Afterwards we toasted marshmallows over the embers. The first time we did this as a family was in my father's hut high up in the Hottentot Mountains in southern Africa. Younger Boy was still a baby at the time and Older Boy only dimly remembers his first barbecue, although he does remember the mongoose that played around the hut in the mornings. But toasting marshmallows has now become a ritual without which any barbecue, or 'braai' as the South Africans call them, is incomplete. The boys hunt for sticks to toast them on – three each with a 'raw' one for afters. Later that evening I cleared away the barbecue and the lone blackbird sang sweetly from the ivy-covered telegraph pole.

Leaving the house to head back up to work, I noticed that the white-blossom relay is on its last leg in the hedgerows – the baton has been passed to the elderflower. In the orchard, flowering starts with the cherry, followed by the plum, then the pear and finally the apple trees. That is all done now, but in the hedge there are white blooms from mid-March to late June. The blackthorn opens first then the blossom unfolds in waves as the cow parsley, the whitethorn and lastly the elderflower bloom. This last bush is very productive: the Beloved has made cordial from its flowers and tonic from its berries, both of which have won prizes at the county show in August. We will have to start thinking soon about our entries for this year. I passed the playground, which was busy and thriving, and left the village. There was a slight haze in the warm air and the edges of the green hills, the blue sky and the white clouds were blurred. The last lighting of the stove can, of course, only be nominated retrospectively because the weather always keeps you guessing, but I suspect we won't be using it again until summer's past. That, at least, is one less job to deal with.

Back in the heart of the city's hum, the Beloved's sister invited me over for supper one evening after rehearsals. Her leg is nearly mended and, to prove it, she cooked up a storm. We ate huge amounts of food and drank similar amounts of wine and rounded off an enjoyable meal with a rousing political argument. It was dusk as I walked back to catch my tram. I stopped to have a beer and a cigar outside a pub overlooking the canal – I thought I needed just that little bit more. I felt an explorer's obligation to taste it all and not miss an opportunity to partake of the metropolitan glut. Actually, I was being greedy. In the country, the Beloved usually restrains my inner glutton. Out of range of her disapproving stare, I sipped my beer and puffed my cigar like a suave *flâneur*. Then I felt a little queasy and had to acknowledge that my capacity for hedonism has diminished. Age, probably, but maybe also because I am these days more used to the cleanliness of simpler pleasures. I waited till the slight wave of nausea had passed, then I raised my glass in silent acknowledgement that she is always right and forced down the last, long gulp. Well, there's no point in wasting it.

It was dark when I weaved away from the pub, but the pavement along the side of the deep, dark canal still radiated summer warmth. Thick bulrushes lined either bank and orange streetlamps flickered on the water in rippled bands. I got to my stop and climbed the steps up to the platform. From my raised viewpoint – a bridge spanning water and tarmac – pissed and waiting for a tram, I surveyed the city at night. A fat, cheesy moon hung above the darkening rooftops. At street level, the strident noise and red lights of traffic dominated. Pearly white lights along the pathway at the water's edge illuminated a pair of young lovers. They were oblivious to the joggers and power walkers who passed them – mind you, the joggers and walkers were equally self-absorbed. Everyone was wearing headphones. They were all trying to dampen the cumulative, overwhelming noise of the city with their own personal soundtracks. With the fogged clarity of alcohol-induced insight, I saw an alienated, fragmented, urban humanity, divided not united by technology – all listening to an artificial, private reality whilst ignoring the shared, actual one … what a loss it was for them! A woman standing near me moved away – perhaps I had made this observation out loud. Then a tram stopped. I lurched aboard and grinned at my fellow passengers, who all averted their eyes.

The play is shaping up nicely. All of the cast have been digging deeper into the text: adding layers to their characters whilst simultaneously focusing tightly on their individual journeys through the play – their 'through lines'. The stage-management team have also been excavating. I begin the play surrounded by worn family volumes and faded estate ledgers from the old house's library. Every time I open one of these 'prop' books, I discover that they have all been used in previous productions of other plays by the same writer. This is immensely pleasing for two reasons. I love that, instead of just dressing the set with any old books that look about right, someone has burrowed away in the theatre's subterranean props room to find the precisely perfect aged tomes. And then, on top of this, the delight I find in turning a page in rehearsal and making this discovery is a genuine pleasure which I can use and incorporate into my actual performance as a scholar looking for, and occasionally finding, useful material in dusty old documents.

This week we ran the play from start to finish in the rehearsal room for the artistic director and various heads of department. It went well. These 'producer's runs' can be good and bad things. When you have people to play to for the first time, you feel their attention sweeping around like a searchlight, highlighting the points of interest. This is useful and salutary. Everyone's game is raised. There is a communal jolt of adrenalin and a general sharpening of intention. On the other hand, the democracy and equality of a well-run rehearsal room fade under the gaze of an audience. The key parts emerge in relief and the lesser roles retreat a little into the shadows. I would be a liar if I didn't acknowledge a little sense of dwindling self as the run progressed. Despite all my work, and indeed the director's attention, my role doesn't feel that rich after all. Maybe everybody feels the same, or maybe I am not as wonderful as I thought.

That Saturday afternoon it was good to be heading home. I was to be the justified centre of attention in my family, if not on-stage – it would be Father's Day the next day, and I had been told the boys had plans. As I drove the sun was higher and the shadows shorter. On the quiet roads, buttercups, daisies and long grass flailed the side of the car. The hedgerows were thickening with honeysuckle, briar rose and fuchsia. The hills were voluptuous. One in particular, which embraces the nearby larger village in succulent green folds, looked like nothing so much as the naked hip and thigh of a recumbent woman.

Early the next morning I was told to stay in bed. Normally I would happily succumb to sleep for an extra hour or two, but this lie-in was enforced. I could see a beautiful day through the crack in the shutters. I wanted to get up and go outside but couldn't. I lay there thinking about mowing, weeding, planting and laundry. I heard clattering from the kitchen; the front door opening and closing; creaks on the stairs; and finally the boys' urgent whispers from outside the bedroom door. Then they paraded in carrying homemade cards and presents and some freshly picked flowers, followed by their mother bearing a laden breakfast tray. The flowers were the first of the pink roses that grow up the side of the piggery; and the presents were loom bands the boys had made themselves – multicoloured bracelets fashioned from interwoven rubber bands. I love them and wear them all the time. The breakfast was tea, poached eggs and bacon, toasted home-baked bread and homemade marmalade. Breakfast in bed is a pleasure more enjoyed in the anticipation than the actuality. I appreciated the gesture but would much rather have come downstairs to eat. I find balancing a tray on my extended legs, whilst sitting bolt upright in order to sip tea and eat eggs without dripping yolk everywhere, quite an ordeal – especially with my family lined up watching me, smiling and nodding at how much I must be enjoying this 'treat'.

I thanked them all heartily for the presents, cards and breakfast and was finally allowed to get up. I looked out the window over the vegetable garden. The day was glorious, the potatoes were flourishing and the lines of sweet corn which I had delegated the Beloved to plant out during the week were only slightly crooked. I remarked as much and instantly regretted it. She made a weird howling noise, which was a mixture of fury and derisive laughter, followed by a scornful, triumphant shout of 'I knew it!' I realised that perhaps I should have thanked her for the work and not noticed any irregularity, or at least not commented on it.

Whilst I have been away, she has been understudying the parts of Mellors and Perkins. She feels she gives a more than creditable interpretation of the former but she is not terribly interested in playing the latter. I dared not say anything about weeding and I decided not to point out that I have black socks and I have very, very, very dark-blue socks and that it is very easy to mistake one for the other and put them away in mismatched pairs. In an attempt to regain lost ground, I suggested we postpone any household tasks and head for the beach. The boys' jubilation drowned out her residual grumbling. I made sandwiches and she fried sausages whilst the boys dug out their swimming togs and wetsuits. Before we left, I rang my dad to wish him a happy Father's Day. He was pleased to get my call but was not otherwise cheerful. Bobba has not been well for some time now, and she is not likely to get better. He is depressed, she is angry and they are both frightened. He spoke to both his grandsons and when I got the phone back I could hear the crack in his voice. There are times more than others, I think, when he is mindful of the price he paid in moving to the other end of the world.

A month or so ago, when we last spoke, he was in flying form. The two of them were in Amsterdam for a few days after a river cruise up the Rhine. They had smoked grass and then gone for dinner. They were politely but firmly asked to leave the restaurant because their stoned cackles were disturbing the other diners. He was still high when he told me the story on the phone from their hotel room. While we were talking I heard a thump, a pause and then a combined burst of hilarity from the pair of them. Apparently Bobba had just rolled off the bed. I reminded him of this and he chuckled, then we said goodbye.

I went into the garage to get Older Boy's bodyboard and noticed that five swallow chicks have fledged. They were on the handlebars of the bike that neither the Beloved nor I ever use. It is hung from the rafters against the wall where they nest. They were perched in a line a few inches from where they hatched – they must be getting ready to fly.

We left the house and drove through the village past the busy playground. This has been a big success. The only complaints have been from one or two people who live just across the road from the green. It seems the zip wire and the swings have proved irresistible to noisy lads leaving the neighbouring pub after a skinful of pints. That is a novelty that will wear off. We drove up and over the hill on the other side of the village and past the old aristocratic estate. Wild purple rhododendron had re-seeded among the hedgerows and more purple was showing as the foxgloves in the verges opened. After half an hour of driving over the rolling hills, we crested a high one and got our first view of the sea – a hazy but sparkling blue band on the horizon. This is a moment that always reminds me of the excitement of such glimpses from my own childhood. In fact, the green hills and sandy beaches are pretty much the same. We head to a stretch of coastline only a few miles south of the place my parents used to favour when I was a boy in Dublin.

We parked at our usual spot and unloaded the car. It doesn't take as much coercion as it used to get the boys to carry something and they are both becoming a little more co-operative. We threaded our way down a steep pebble-strewn path to the sandy patch between two rocky headlands. A little cove between two much longer, open beaches, we usually have it to ourselves. We spread our rugs and put up our folding chairs and the

large sunshade whilst the boys writhed into their wetsuits and plunged into the water to splash, paddle and cavort. The Beloved stretched out to sunbathe and I poked around in the rock pools looking for crabs. The boys joined me. We only found a couple of small live ones, but there were an awful lot of dead ones about, or bits of ones. Where the sand adjoined the rocks was a crab graveyard. We collected assorted claws, legs and pieces of shell and piled them together in a heap, then Younger Boy set to. After a while he called me over to see his 'crab creation'. Using all the fragments we had found, he had fashioned a kind of crustacean Frankenstein's monster – a model of a giant mongrel crab. It was pretty cool.

We had lunch. The salty air seasoned the cold sausages and warm beer and made them delicious despite the grittiness of the ubiquitous sand. Then everybody returned to their respective tasks: the Beloved to lying on the rug and the boys to playing.

I opened another beer and looked around. My appreciative eye lingered on her supple, supine body glowing in the sunshine. I glanced over to the boys. The older one was tentatively surfing on his bodyboard. Then he stopped and bobbed idly in the waves, smiling at nothing in particular. The younger one was lying on the beach opening and closing his legs and sweeping his arms up and down – he was making a sand angel. Then he also stopped and just lay there, in the shape he had scuffed for himself, basking in the sun. Neither of them had any cares. They were both lounging in the ample width of the present moment. Each of them self-sufficient at the centre of their own universe – I took my cue from them and sat silently. The sun moved slowly westward and the waves lapped softly. My mind was as still as my body. Thought seemed superfluous and too much effort. The play was forgotten. I was aware only of where I was and whom I was with.

We left when the time was right and stopped for ice-cream on the way home. I ordered and paid for four ninety-nines – small ones. The boys grinned at the shopkeeper as they asked politely for strawberry and chocolate syrup, and they emerged with giant cones. When we got back, we rinsed and hung out the wetsuits and swimming togs. In the kitchen garden, the haulms of the early spuds were getting high so I earthed them up because it needed to be done. Then I did some light weeding amongst the peas and beans just to drop the Beloved a little hint.

Back in the theatre, the technical rehearsals have progressed with uncommon ease. This is a part of the process when nerves can get frayed. The actors are getting used to walking the set for the first time and occasionally encountering obstacles undreamt of in the rehearsal room. The technical crew are intent on working through the lighting and sound cues and are not hugely interested in the niceties of performance. The stage management are charting the movement of all props and stage furniture and ensuring that they start and finish the show in the right places. Up in the box, the stage director is noting and timing everything so that, over the backstage loudspeakers or the crew's headphones, she can 'call the show'. That means cuing every single actor's entrance; every single lighting or sound cue; every single scene change and the fading up or down of the houselights in the auditorium. During this slow process, sitting out front and overseeing the needs of cast, crew and wardrobe, as well as accommodating the vision of set, costume and lighting designers, is the director. He has to take his production completely to pieces and then put it back together with the full panoply of theatrical artifice in order to give the play full, bold life. This takes skill, patience and a calm environment. Fortunately we have had all three: the facilities of the national theatre gave us the time we needed and the expertise of the best technicians in the land. And as I have discovered before with this director, the transition from the rehearsal space to the playing space happens with a smooth, anxiety-free elegance. There was nothing to worry about and nothing was going to go wrong. Except, of course, something did – Venerable Balding Actor had to pull out of the show.

We were all aware during rehearsals that he had health issues and that his usual vigour had been compromised, but he is held in high esteem and great affection by cast, crew, director and producer, so everything possible had been done to husband his energy and help him deliver what was set to be a delightful, quizzical performance as the family's eccentric, elderly uncle. The biggest drain on his reserves had been self-inflicted. He fretted and worried that he might let the rest of us down. I understand this acutely – some years ago I had to pull out of a production due to ill health. I knew how he felt. During the dress rehearsals I kept him fed and watered. I was sharing a dressing room with him and Venerable Bearded Actor, and their reminiscences of theatrical triumphs and cock-ups from well back in the last century were a joy to listen to. At the first preview performance he pulled it out of the bag and delivered in full. The house was packed and appreciative and the play soared. But ahead of us, we had a long run through what promises to be an unusually hot summer. Another venerable actor, who is both balding and bearded, had been on standby for a week and he was going to take over – the decision had been made by management. The fact that it was the right one was confirmed when the man himself appeared in the bar after his performance looking dignified, mightily relieved and ten years younger. The rest of us were nonetheless a little shocked and quite upset. There were handshakes, hugs and a few tears. He was entirely upbeat – assuring us that he would not be going away, that he would still be in the theatre right through to the opening and beyond, and also that he would be available should his replacement need to be replaced. The Venerable Bald and Bearded Actor who has joined us played, it transpires, my role when the Abbey last revived this play – another layer and link in the succession.

Clapping Crowds

As I walked through the bright whirl and din of the city on Midsummer's Day, I kept an eye open for anything that might have been an acknowledgement of the summer solstice. I spotted a group of lithe young people dressed head to toe in scarlet Lycra. They were dancing in celebration of something so I investigated. They were ritually honouring the opening of a new cut-price clothing store. City celebrations of the solstice used to be utterly raucous affairs. Midsummer gatherings led to much drunkenness and debauchery, unsurprisingly. So, unsurprisingly, the city authorities and the clergy combined to suppress them. In 1710 an act was passed banning 'dangerous, tumultuous and unlawful assemblies'. The young ones these days, who get riotously trashed in certain quarters of the city on a Saturday night, could argue that they are merely reconnecting with the ancient festive traditions of their peasant forefathers.

The summer solstice is the longest day of the year. The word has Latin roots – *sol* meaning sun and *sistere* meaning to stand still. In Celtic Ireland it was yet another fire festival. Communal bonfires were lit once more and again became the focal point for dancing, music and general roistering. Improvised fireworks were common – boys competitively threw burning brands from the fire high in the air and people cheered and went 'Ooooh!' as they watched the trail of sparks. The Celts obviously loved standing around big fires and were consistent if not inventive at justifying why they ought to. The Midsummer reasons were the usual ones: good health, good luck and good magic – both communal and individual. But to avoid absolute repetition, small, specific refinements were invented for this solstice. You could invoke the power of the flames to forestall a bad harvest or to prevent the potatoes from being blighted – both of which were crucial. Immersing yourself bodily in the smoke was a protection against witchcraft. But the specific power of this particular festival was in disposing of troubles – of actually incinerating them. Especially bothersome weeds were burnt to prevent them recurring and bad-luck charms could also be got rid of – any artefact that was thought to have supernatural powers was only safely destroyed by being burnt on the Midsummer fire. In the nineteenth century, Sir William Wilde, Oscar's father, recorded the many dividends people claimed were the benefits of jumping over a bonfire at Midsummer. These beliefs were common countrywide thirteen hundred years after Saint Patrick, which suggests that Christianity has been an enduring but thin veneer in Ireland.

And sure enough, as I discover more about the old Irish fire festivals, with their heady mix of fierce celebration and vital magic, I detect the lingering presence of a goddess. The solstice was sacred to a deity who I think has become my favourite. Her name was Áine. Mythological accounts of her origins vary slightly: she was either the wife or daughter of the sea god Manannán mac Lir. She was often personified as a red mare, and whilst she didn't have as many attributes as Brigit, the few she had are ones I am quite keen on: summer, sovereignty, wealth and love. She also kept a benign eye on cattle (who didn't?), and on the Midsummer solstice, torches lit from the sacred fires would be brought among the herd to ensure her protection – after all, a fire festival wouldn't be complete if the cows didn't get to join in. The sovereignty Áine symbolised was intimate and personal. She embodied self-determination and helped people claim individual power over their own destiny. It was Áine you invoked in order to experience true joy in life. The wealth she denoted was the largesse of her summer season: the fat cattle in the rich fields, the ripening corn, the full-grown lambs and the rivers full of fish.

But it was as goddess of love that I most esteem her. Áine's love was not an ascetic, anodyne, spiritual affair, but a deep, earthy, erotic joy. She had both appetite and aptitude. She is to tantric sex what white-water rafting is to a paddling pool. She often took mortal men as her lovers and taught them how to properly, joyously satisfy a woman. If I had known of her when a hormonally explosive teenager, I would have become her acolyte. I would have built mighty bonfires in her honour and prayed to her to come to me and ravish me in a sacred manner. There are many legends of the mortal men to whom she bore strange, ethereal children, one of whom was known as 'the Magician', and scholars speculate that this may have been the origin of the character Welsh Celts knew as Merlin.

My own solstice had many attributes. In the morning I mowed the grass and fixed a pair of garden benches for Fond Aunt. Then the Beloved rang and asked me to get sun lotion for the boys and netting to keep cabbage white butterflies off her cauliflowers and broccoli. I didn't question the use of the pronoun 'her' – she has taken up the horticultural reins to a large extent recently. She also mentioned that the Farmer Friends were coming over for supper that evening to vaguely celebrate Midsummer's Night, and depending what time they finished making hay, and therefore what time they ate, I might see them later when I got home for the weekend after the show. In the afternoon I had a swim and a sauna, watched a bit of telly and went in to the theatre for the third preview. It went well. I drove home after the show delighted with life. About half an hour from home a guard stopped me. There had been an accident up ahead – thankfully, not fatal, but the road was blocked and I would have to turn back. He gave me rough directions for an obscure route over small back roads. I roughly followed them, soon got lost and trusted to Áine. I eventually found myself on a familiar road much closer to home than should have been possible. I had somehow magically bypassed a small town in less time than my usual route would have taken. I still don't know how I did it.

When I walked into the kitchen, they were all still in conference around the table. The Farmer Friends had been quenching the deep thirsts acquired by baling hay all day and the Beloved had been gladdening her heart with much white wine. They had saved me a plate of food and there was plenty left to drink, so I quickly got up to speed. The Beloved was swearing a lot. After being alone with the boys all week, she felt the need to drop an expletive or two. The Farmer Friends found this hilarious – because of the wit or the wine, I am not sure. I took advantage of the levity to mention the subtle distinction between black and very-dark-blue socks and said no more. I risked it because, in the first place, she was cheerfully pissed and we were in company, and I also figured that socks wouldn't particularly hurt if she rammed some up my arse.

The next morning the sun rose on yet another gorgeous June day. Officially, the high point of summer has now passed, but it feels like it is only getting started. In the field across the road you can see the result of the Farmer Friends' recent work. The giant golden lozenges of the bales stand in ranks down into the valley. They literally made hay whilst the sun shone. It has been perfect weather for such a task: for five consecutive fine days after it was cut, the hay has been drying in the sun, and now it has been gathered into the familiar fat yellow cylinders, like thick slices of a giant Swiss roll. According to the forecasts this beautiful clear spell will last, so the bales should get the optimal few days seasoning outdoors before being brought into the barns for winter fodder. I spent part of the day catching up with various jobs – I mowed, I weeded and I laundered. These small tasks are usually undertaken by Mellors and Perkins, but I rarely feel their presence these days. I don't have as much time as I used to – I deal with chores in a much more perfunctory manner and I tend to forget my little gallery of supporting players. Now that I am playing an actual part, these characters seem less amusing and I think I will lay them to rest. I have relinquished the invented roles to play a real one.

The remainder of the day I largely spent just enjoying where we live. There is now a second pear on the second tree, the cherries are dwindling so there may not be much fruit, if any, from there, but the plums should be plentiful. The roses are smothering the wall of the piggery, and at their feet the Beloved's herb bed is bursting with chives, marjoram, oregano, parsley, sage, rosemary and thyme. In the kitchen garden, the early spuds are flowering – we should be eating them before too long; the sweet corn is bedding in; the salad leaves are ready; and the peas and beans are climbing ever higher. In the hedgerow the elderflower is currently king.

In the late afternoon we barbecued the last of Milo's shoulders. The Beloved's mother joined us and, in anticipation of a forthcoming international soccer tournament, the boys insisted we play until the early evening. Their interpretation of the rules is elastic and self-serving. The adults retired from the field thoroughly beaten. I had a restorative beer and sat in the lengthening shadow of the big cherry tree. Pigeons were cooing, the lone blackbird was hopping beneath the buddleia and the swallows were giving an aerial display. The gaps between the tall trees at the bottom of the acre opened up extended views away to the far hills: the slanting sunlight glanced off their summits and the long shadows in the valleys between were warm and blue. The bank at the front of the house was getting hairy and the taller weeds in the verges of the hedge around the field were looking belligerent. I needed to borrow a heavy-duty strimmer. Then I thought, 'Ach! That can wait.'

The morning after the longest day of the year dawned with a Mediterranean blaze. I was briefly dazzled when I opened our bedroom shutters. The cloudless sky was azure and the vegetable garden was green and green and green. The foliage of the peas, beans, potatoes, courgettes, carrots, lettuce, leeks – any and every plant – is a subtle but distinct shade of that hopeful colour. I got the boys their breakfast, made tea for me and the Beloved and took a mug outside. In the hedgerow across the road, the red, juicy blobs of the fuchsia flowers were dripping like carmine tears from their drooping stems and the burnt-orange sprays of montbretia were starting to emerge. In the field behind us Male Farmer Friend was putting a couple of late calves out to grass.

Most of his herd are mixed breed. This gives them what he calls 'hybrid vigour'. The cattle are healthy, fertile and strong. The bull is brought to the cows at a particular time: insemination is swiftly achieved and calving presents few problems. This means that the whole process can be pretty accurately timed so that the young, newly weaned calves are brought to pasture when the grass is ready – usually late April or early May. He also has a few pedigree Parthenaise cattle. These are a lovely-looking French breed, golden like an impala but ten times the size, with huge rumps which just beg you to sink your teeth into them. They are valuable but, being pure bred, they bring various problems, one of which is unpredictable and often late fertility. There can be delays, and every year there are a few straggling calves who are put out to grass when the summer is well advanced.

I didn't have time to straggle. I finished my tea, put on some work clothes and did some strimming. In centuries past, Irish migrant workers left home after the solstice to go and work the harvests in England and Scotland, but before they left, they would save the turf and earth up the spuds. We don't have a turf bog but we do have potatoes, so I earthed up the main crop. Using a hoe or spade, you draw soil up and around the plants to make a raised mound from which the greenery – or haulm – pokes up. This encourages more tubers to grow. When I finished, I had a cup of coffee under the cherry tree with the Beloved and mentioned I had put the netting for the broccoli in the garage. Then I kissed everybody goodbye and drove up to town for the last two previews before the opening night.

The morning of the opening dawned gloriously and I went for a swim in the leisure centre twenty minutes' walk from Fond Aunt's house. I strolled past chattering shopkeepers as they pulled down awnings to shade the sun-baked pavements. I spent about an hour in the pool, interspersed with short sessions in the sauna and steam room. Feeling thoroughly relaxed, I caught a tram into the city. I wandered into the National Gallery and bought some nice cards for good luck wishes. Then I cut through Trinity College and walked alongside the playing fields. A cricket match was in progress so I stopped to watch for a moment. The traffic noise receded into the background. Someone bowled, someone swung and there was a slight delay before I heard the 'tock' of the leather ball on the willow bat. It was a mighty stroke. The spectators applauded, the ball flew and someone pelted after it. It all looked a little strenuous so I walked on.

In the theatre we met on the stage and ran a short section of the play to tune in to each other and then the director made a brief valedictory speech, as he usually does. He spoke eloquently about the play and the work we had all done on it together. He wished us all good luck, thanked us for our efforts and arranged to see us in the bar afterwards. Then he disappeared. I felt a little sad – partly for us and the fact that our leader was saying goodbye, but also for him: he was letting go of a project he had been working on for a lot longer than any of us. What he was doing, of course, was relinquishing power and handing it over to us. It takes a strong, self-confident director to do this. It's ours now.

For the next hour or two, the theatre was abuzz with the usual first-night excitement and hyperactivity. Most actors will tell you that it is all a little absurd – why should one particular night be 'the' night? It's silly – best get it over with and then settle down to do the job and play the play. I agree … and yet I am still a bit of a sucker for the special occasion. Cards, gifts, flowers and – a new one this – muffins circulate. There is a tangible excitement backstage. The warm-up on-stage is a little giddier than usual, with more ululating than normal. Not everyone does this – the old lags in any cast tend not to bother with voice exercises before a show. I am on the border between the two camps in this regard. I sort of nod in agreement with remarks such as 'Sure you'd bollix your larynx with antics like that,' as the dressing room echoes to a virtuoso vocal swoop over the backstage speaker system. But then I sneak off and wander onto the stage for a bit of a sing. It does clear the tubes, and to be honest I like to hear my voice resound around the space. I am quite sure that, meanwhile, in some dressing room someone is saying, 'Listen to your man … thinks he's Count John McCormack.'

Just before the half-hour call my phone buzzed. The Beloved had sent a picture of the boys grinning through armfuls of wildflowers picked from the bank – a digital first-night card. Then she rang to wish me good luck. She was lying in the hammock, watching the newly fledged swallows performing aerobatics above the house. That sounded nice. I was sweating profusely into the woollen vest that wardrobe had given me. An accurate character detail, no doubt, but as it was invisible under my crisp button-collar shirt and neatly tailored houndstooth checked jacket, I decided I could play the part without it.

The show was relaxed but sharp and the response was great. When the applause had died away, we were herded up to the top floor of the building to meet the lord mayor of the city and the president of the country. This was an exciting and signal honour. When I returned to Ireland nearly twenty years ago, I was proud of the vision and dynamism of our then president Mary Robinson and thrilled to discover that the government's presiding minister for the arts was a poet. This same man now inhabits Áras an Uachtaráin – the presidential mansion in the Phoenix Park. I shook his hand and told him of Younger Boy's amusing mispronunciation of his name. He feigned delight and chortled graciously. In the bar afterwards I shook many more hands and accepted many congratulations, most of which seemed genuine. I think we have a success. The bar seethed noisily and amusingly for a couple of hours and then a few of us headed round the corner to the nearest late-opening pub for more unnecessary drink. The last tram had long since gone when I left. I hailed a cab and quizzed the driver at length about his life and loves. When he dropped me at Fond Aunt's house, I gave him a very large tip.

When I rang home the following afternoon, my account of the night was upbeat but vague, and very softly spoken. The boys' accounts of their own successes were much shriller. Their school reports were excellent and they had both acquitted themselves honourably at the sports day. Younger Boy was piercingly loud. He yelled that he and his classmate – a little girl he seems especially fond of – had won the three-legged race. He also announced proudly that she had done equally well in the tag rugby. He seems smitten with her. It is wonderful to be up in town, performing in a successful production at the national theatre and meeting a president I admire, but I am missing small delights. However, one pleasure the country cannot offer is dropping in to see a lunchtime show. On the last Saturday in June, I saw an hour-long two-hander set during a course of evening dance classes. The two charming, funny and versatile actors tangoed, waltzed and foxtrotted in a simple, joyous, humane and lovely show. Dancing really should be a required part of adult life. It should be a civic duty, along with voting, paying taxes and not kicking traffic wardens. More than anything, the piece was irrepressible fun and the day ended in the same mood.

That evening, a clatter of aunties and a lone, brave uncle came to the show. Fond Aunt is a regular theatregoer. She had seen the original production of the play and was keen to see what we had done with it. She brought along a brother, a sister and two sisters-in-law. They were all out for the night and ready for divilment. They thoroughly enjoyed the performance, but the play was really only a prelude to the evening's entertainment.

They live in scattered parts of the city and they like an excuse to get together. When I got to the bar after the performance, I followed the sound of their laughter to the corner they had commandeered. Drinks had been bought and more were to follow. I settled easily into their collective congratulatory gaze and swigged the pint they had ordered for me. I wasn't allowed to relax for long. There are some attractive young hunks in this cast, one of whom is well known from a couple of TV dramas. The aunties thrust some programmes into my hands, told me to 'go and get that nice young man's autograph' and then shoved me in his direction. I complied. Then I was allowed to finish my pint and was even bought another. Fond Aunt noticed Venerable Bearded Actor at the other end of the room and she reminisced about many fine performances she had seen him give over the years. She mentioned an especially vivid memory of him wearing thigh-high leather boots in something or other decades ago. There was much raucous chat and more loud laughter. The aunties waved and went 'yoo-hoo' as the Hunk departed, and we were the last to leave the bar.

I suggested another drink at the late pub around the corner. They unhesitatingly agreed. When we entered the pub and saw some of the cast amongst the other customers – including Venerable Bearded Actor and the Hunk – they were delighted. They swiftly occupied another corner and I bought a round. Earlier, when I had asked the Hunk to sign the programmes, I had mentioned that the aunties were giving him the glad eye. He now came over to where we were sitting and gave them the full attention of his easy charm. They leaned forward in their seats and nodded, giggled and chattered. He leaned down towards them and blushed, beamed and nattered right back at them. After substantially more minutes' talk than duty required, he thanked them for coming to the show, wished them a good night and wandered back to his pint. There was a brief silence. I felt rather than heard a collective sigh and then they all started speaking at once. Apart from Fond Aunt, that is – she hadn't been swooning quite as much as the others had been. She regarded the Hunk with a doubtlessly benign eye, but her real attention was on Venerable Bearded Actor, and she once again murmured something about high leather boots.

I went over to where he was standing and asked him to join us for a moment so I could introduce her to him. I mentioned the boot part and he knew exactly what show she meant. He strolled elegantly over to our table and they shook hands. She told him of her great admiration for his work over the years and he thanked her graciously. Then, referring to the boots, he said, 'I believe you saw that show,' and grinned wolfishly. She smiled her agreement and her eyes glinted. Perhaps I have misnamed her – maybe I should refer to her as Randy Auntie. We left the pub not long after. There was a cloudburst – one of those sudden, heavy summer rains. We were drenched as we walked to the tram stop. Fond Aunt didn't seem to notice. I did but wasn't that bothered – the kitchen garden would benefit.

Racing Ducks

According to the cliché, city life is fast and country life is slow. For most of July I have found the opposite to be the case. Up here in town I have been playing the same part nightly, and during the day I have settled into a loose routine. I have a swim and a sauna, I write some more of this book and each successive evening I go to the theatre to inhabit the same play. My performance has subtly developed as the run has progressed – tiny, incremental changes of emphasis have made my characterisation more complex and nuanced. Or so I like to think – it is probable that these miniscule shifts of tone are quite invisible to anyone else. This quantum psychology keeps me interested as night after night I re-enter the familiar world, static and unchanging. The passing weeks have merged into one long, languid August afternoon on the lawn of an old Georgian mansion.

Meanwhile, my brief weekends down on the acre have been hectic. I work frenetically to assuage the guilt I can no longer ignore at the extra pressure the Beloved is under. She needs to harvest, pickle and freeze the produce of the vegetable garden before it rots – the growth has been prolific and profuse. There is daily increase, never mind weekly, and each time I come home I see distinct differences, not just in the height of the peas or beans, but also in the texture of the trees and hedges, or in the colours of the fields on Sliabh Buí. I see the country in a series of snapshots and the town in one lingering image. Rural life is lit by stroboscope and living in the city has the tranquil tone of a still photograph.

The two worlds overlapped when the Beloved brought the Farmer Friends to see the play. They drove up and down from the village for the evening. I joined them for an early dinner just across the river from the theatre and waved them off after a quick drink in the bar following the show. The Farmer Friends were effusive and apparently discussed the play for most of the drive home. I didn't really talk to the Beloved much, as she was table-hopping around the bar, catching up with other cast members who were obviously much more interesting. We talked at length by phone the next day. She had loved the play and made complimentary comments about my performance. I was pleased – I value her good opinion. She is biased, of course, but her judgement is acute and she is always honest. When appropriate, she is also critical. My work has sometimes bored her and she has told me … twice … I remember both occasions with pursed-lipped clarity.

She was not bored this time but the boys would have been, so when they came to town we didn't bring them to a performance but I gave them a tour of the theatre. We began in the rehearsal room, where work has started on the next production, and made our way down from there. They loved the view from the high windows across the city. In the control room, amongst the lighting and sound desks, they had the same sense of height – from just below the lighting grid and at the back of the auditorium, they looked across the empty space to the distant stage. They were allowed to sit at the desks and twiddle some knobs, which they declared to be 'cool'. Then we went through wardrobe, hair and make-up – full of wigs, frocks and lipstick, which merited merely a cursory nod. Down the stairs past the dressing rooms, along corridors hung with the Abbey's collection of theatrical portraits and century-old costume designs rarely seen by the public, we finally emerged onto the stage. They were fascinated by the fake grass which had looked entirely convincing from the control room. They skipped about on it like lambs in the green fields. I asked someone to bring up a couple of different lighting states so that they could feel and see for themselves the dazzle and heat. In my dressing room they took it in turns to sit in my chair, rooting in my make-up box and looking in the mirror surrounded by white bulbs – just like you see in the movies!

The high point for them was the highest place yet. One of the crew took us up into the fly tower. Way up above the stage, it is where lighting bars and long black drapes – or 'legs' – and, indeed, the curtain are housed. From there they can be lowered or hoisted out of sight as required. There are also cylinders attached to pipes which stretch from wing to wing to provide rain or smoke if needed. The boys thought that this was utterly excellent and asked a torrent of questions about bangs and flashes and, of course, blood. When they got to pull on a rope or two they took it very seriously – intent eyes shining from attentive faces. Back at stage level in the darkness of the wings, behind the wooden flats, their sight acclimatised to the comparative gloom and they noticed a large room-sized metal cage. I asked for the padlocked door to be opened. They entered and stood in silent, reverential awe in the crew's capacious tool store. We were about to leave by the stage door when one of the stage-management team stopped us and asked if we would like to see the props room. She took us down to a subterranean corridor beneath the stage and unlocked a shabby, nondescript door. It opened into a long, low room I had never seen before. The walls were lined with floor-to-ceiling shelves packed with bric-a-brac from countless previous productions: swords, luggage and candlesticks; crockery, guns and walking sticks; antique telephones, golf clubs and decanters – I spotted a silver hip flask I had used in a Wilde play years ago. In one corner there was a heavenly host of devotional statues: angels, virgins and a congregation of plaster saints – mundane mass-produced symbols which had been used in a theatrical artifice to ritually evoke a notional god. They would most likely have been acquired from the very religious shops the faithful would have used to purchase their icons – the same figures bought to be prayed to or played with.

I was pleased and proud to show them around – proud of my family and proud of my work, and pleased that I was able to facilitate a memorable moment for us all. We have spent a lot of the summer apart and I regret it. The Beloved has sent me pictures, phone to phone, of the trips I have missed as they unfolded. It has been great to see images of the various jaunts in real time. But it also emphasises the distance. In one photo they are leaping in the crashing surf on a beach in Cork. Another, in a fairground, shows them careering around in dodgems – each intent on taking out the other. And there is one of Younger Boy in a canoe on a river. He is in his wetsuit, wearing a life jacket and plying a paddle like an Amazon explorer: his mouth grimly set; his forehead a knot of concentration; and his eyes gazing intently forward as he scans the peat-brown water. God, I love him. I love them all and I miss them.

Home on the acre, I missed the first courgette flowers opening in their glowing yellow starburst; I missed the first picking of peas, crisp and sweet from the pod; and I missed the first storing up for the colder months – the pickling of the shallots. Again, the Beloved sent me a picture of Younger Boy carefully pouring the spiced vinegar into a giant jar of them. He was eager to help, as he and his brother adore eating them. We had only recently finished the final jar of last year's crop, which I thought delicious. However the boys felt the pickling mix was a touch too sweet – they both preferred a slightly sharper bite. I hope the Beloved took note because I wasn't there to do so.

When I am not professionally employed, I hunger for work. Now that I have some, I resent the fact that it keeps me away from home. I perpetually feel that I'm missing something somewhere. Why can't I have it all? Actually, right now I do. My desires are being satisfied in every area of life, there just doesn't seem to be enough of me to absorb and enjoy it all. It's all too much to handle, which is extremely annoying – life is flawed even when it's perfect.

One major event in the village's calendar I didn't miss was the juvenile football club's second annual duck race. As vice-chairman, I knew my presence was required on the Fair Green for the festivities. I was anxious that in my absence I had been allocated a task that no one else wanted – something beyond my capabilities. It was a relief to discover that after my gross numerical incompetence last year I was to be kept well away from anything requiring counting. When I turned up on the appointed Sunday lunchtime, I was stuck at a far corner of the field and told to supervise a bucket of sand. Stuck in the sand was a forest of lollipop sticks: when you pulled one out it told you whether you had lost your money, got it back or won a hotdog – one euro for three goes. I think I made twelve quid over the course of the day.

At the other end of the field I could hear the Beloved's usually clear tones muffled by the public address system. She had been lassoed by the committee to commentate again, and I heard occasional snatches of her voice as she attempted to garner more contestants for the welly-throwing competition or asked if there was anybody for the last few choc ices. As the shadows lengthened, the sky darkened and the air grew heavy and moist. Rain threatened but never came. Instead a fine, steady drizzle settled in for the afternoon. The kids didn't seem to mind, and when the under-fives football team took to the pitch to play a demonstration game against a nearby town, there was enthusiastic juvenile cheering from the sidelines. It was a bit of a grudge match, and the fact that the players' jerseys hung around their ankles did nothing to dampen the fierce competition. Younger Boy reported later that he heard one of the opposing team refer to someone on the home side as a – he thought for a moment – 'a doer of the F-word!' He grinned as he told me. He was careful not to pronounce the offending syllables but he knew he had conjured the word in my mind.

This had all been by way of a curtain raiser for the main event, the Grand Duck Race, and this year it was nail-biting. Twelve months ago, it was discovered that the river doesn't flow quite fast enough through the village. During the many qualifying heats, the hundreds of yellow plastic ducks had slowly bobbed in a conglomerated mass and dawdled interminably along the length of the course. By the time the final came, even the most determined village wits were exhausted with prolonged jeering. Shouting fake encouragement with mock urgency loses its humour with repetition. Enthusiasm dwindled, the ironic cheers faded and eventually the cries of 'Come on, number three hundred and forty two!' died away. A lesson was learnt. This year two committee members, stripped to their shorts, waist deep in the water and armed with leaf blowers, followed the ducks along the river and wafted them over the finishing line to vigorous applause.

The following weekend I arrived home after the show to an empty house. The boys and their mother were down in the south-west and would not be at home till the morrow. The complete quiet was an abrupt contrast to the city. I went outside and sat for a while with a drink in the dark. Thunder had been forecast and the air was warm and muggy. A bat or two flitted by as I sipped my wine and listened to the silence.

I rose late the next morning and mooched about. The day was still fine – no sign of stormy weather – and the fields of barley in the valley were turning gold. In the garage the swallows had moved to another nest in another corner and were now incubating a second clutch of eggs. They were also crapping on another set of tools. Then I examined the vegetables. The courgettes and cucumbers were showing promise, as were the leeks; but the carrots were poor and the garlic was rubbish. The canes and netting for the peas and beans were a forest of green tendrils and splayed leaves, and the first early potatoes looked ready to eat. All of the beds needed weeding. For a moment I contemplated doing some … she hadn't bothered … why should I … but someone ought to … I let it go and went to borrow the sit-on mower to cut the grass instead. I made coffee and looked at the weather app on my phone which told me that rain wouldn't be with us till late afternoon so I thought we could risk a barbecue. I chopped some logs into small pieces, got some homemade burgers out of the freezer and dug up some of the spuds. Then I drowsed in the hammock until they arrived.

I half-heard the car pull up and then totally heard the loud yells as the boys clattered through the kitchen garden and leapt into the hammock with me. That was fun but not very restful. There was a lot of wriggling and both boys fell out more than once. I clung to my prime position and remained swinging but I was beginning to get dizzy so I suggested kicking a ball around. They agreed and swiftly invented a new game. It is called 'Wrestling Football'. It involves jumping on Dad and then kicking a ball at him. Hard. The rules are esoteric. I failed to understand them but the boys were very clear on when you were allowed to do what.

We had our barbecue in the middle of the afternoon. Eating the potatoes was very nearly a religious experience. Floury, steaming, skin peeling back, rivulets of melting butter – the quintessence of summer – we devoured them as the storm clouds gathered. The sky darkened all around us. Sliabh Buí was invisible under a tall mass of gunmetal blue–black cloud, and so was

the high shoulder of our hill behind us, but we sat in a bright patch of sunlight throughout our meal. It was late afternoon when the thunder broke with a bang. I was in the kitchen garden thinking about weeding again when the rain suddenly spilt from the sky and saved me from that task. Trapped by the cloudburst, I took refuge under the silver birch and watched the bushy potato plants take a battering from the downpour. Even under the tree, I was still getting spattered, so in a lull I decided to make a dash for the house. Halfway there an especially heavy and extremely localised squall drenched me, and just as I got to the front door, the rain stopped as abruptly as it had started. I stood dripping on the doorstep and looked about me. The freshly mown grass had a wet green sparkle and a shining double rainbow straddled the valley. It started beyond the sheltering cattle at the bottom of the field over the road and ended at our lower gate.

Thunder still threatened the following day but held off for most of the morning. I just couldn't let it go, so I did a little light weeding amongst the courgettes and cucumbers and asked the boys to help. They joined me for minutes and then disappeared. I moved on and weeded between the lines of sweet corn – they were planted out perhaps a little later than ideal (and just that little bit crookedly) but they are thickening and climbing nicely. Sweet corn is probably one of the most satisfying vegetables to rear successfully. It needs more work than most others: propagating, bringing on in the polytunnel and then planting out; but one small, wrinkled seed grows to a tall, opulent fullness and produces fat, sweet cobs which we all delight in. I shouted for the boys to come and admire them with me but there was no response. I could hear the older one kicking a ball around at the other end of the acre; I couldn't see where the younger one had got to and then I heard the sound of chomping from nearby. I found him sitting nestled between two rows of climbing peas twice his height, helping himself, pod after pod. I pulled out my phone and took a photo. He didn't notice. He was totally intent on his hand-to-mouth grazing.

Ponderous clouds predominated on the drive back to town and by the time I reached the motorway it was raining steadily. Visibility was poor but traffic was nonetheless fast. Driving slowly causes its own hazards so I gripped the wheel and joined the communal trusting to luck. Wet grey tarmac ahead of me and burnished cornfields behind. Warm and wet weather like this is ideal for transmitting potato blight and I hadn't had a chance to spray the spuds. There just isn't enough time. And there is much yet to do … much left undone … many minor annoyances. There is still no netting over the cauliflowers and the beans were planted late. The Beloved lent the spade to her pal and broke my jig-saw cutting sticks. I recall seeing lots of cucumbers forming, but they seem to have gone and don't appear to have been pickled. It is a mystery. Is she letting them get too big and then giving them all away? Like that year when the huge crop of courgettes grew into marrows and we were chasing departing visitors down the path, forcing them to please take some? She said she would do some weeding but I see no evidence. All I can do is murmur my intention to weed, thereby drawing attention to the fact that she hasn't. She sees me at it and tells me to rest on my day off. I can't! There is far too much to be done. Besides, if I did take it easy, her moral superiority would be unassailable. The rain got thicker. Younger Boy had been tricky to deal with yesterday, and again this morning. He is cross that I am away so much and acts up when I am home. He doesn't like me being gone but gives me a hard time when I'm back. It's very difficult. Why can't he act his age? He is six. It is sometimes harder for me to act mine. I feel oppressed. It's all too much. I need more time to relax with my family. There are hardly enough hours in the day to wrestle boys, cut grass or tie up peas and so forth, let alone find time for the Beloved and I to … well … reacquaint ourselves with each other. We need to make room for that. When the boys were younger, we found children's TV very useful as a means to keep them engrossed whilst we went upstairs. This resulted in some odd Pavlovian associations – for a while, every time I heard the theme tune to *Bob the Builder*, I got an erection. It wasn't always appropriate. No time and little energy this weekend. Some idiot was driving far too close behind me. The windscreen wipers were swishing and sluicing. Huge trucks roared past. The spray from their tyres added to the deluge. My jaw ached from gritted teeth. I was tense, tired and feeling my age.

I hate driving in the rain.

I got back to Fond Aunt's in time for a swim and sauna before the show. This has become a respite and a necessary part of my routine. It is simultaneously relaxing and invigorating. It clears my mind and eases anxiety. After a few lengths, I settle into a soothing rhythm and I emerge recharged. I have searched my conscience for a twinge of culpable guilt at this weekly escape to the city and I haven't really found any. Well, perhaps a little. I did ring home to wonder aloud that if she happened, by any chance, to find herself with a spare half hour sometime, and if she was a little bored, she might just want to spray the spuds … that had been unfair pressure. But there is no benefit in being hard on myself – I am in the city because that is where the work is. I do what I can when I am home and my days in town are hardly lazy: I write, I swim, I act and I also do the odd job for Fond Aunt. This could be anything from grass-cutting and vacuuming to advising on her outfit for an appearance at the Albert Hall. She is a member of a group that meets regularly to do choreographed movement to music. Not quite dancing, it is designed to help sustain suppleness and agility, but it also has a high artistic content. Every five years, her group and scores of others like them from all over these islands meet at that vast Victorian barn in Kensington for a two-day festival. The troupe leader is a woman of conservative tastes who likes to adorn her comrades in simple black leggings and leotards. Fond Aunt has had enough. This year she has managed to assert her moral right to contribute some new costume ideas. She showed me her plans. They involve a lot of sequins and a shimmering purple skirt slit to the thigh. There may be trouble ahead.

She may have her sights set on glory at the Albert Hall – mine are focused much closer to home. This August I am determined to get first prize for chutney in the county show. In previous years I have managed two third prizes and twice I have been 'very highly commended'. This last category feels very patronising – why don't they just call it 'fourth prize'? It carries an implication of 'Well done, dear, very nice … but you don't quite deserve a rosette.' It still rankles. The Beloved thinks I am being petty when I gripe about this 'award'. She can afford to be smugly superior – she has won many first prizes and countless seconds and thirds. She is always among the rosettes. When she achieves the top spot, she walks away from that category and never enters it again. She calls this being 'gracious'. It does mean, though, that her options are dwindling. She has won for raspberry jam, apple and geranium jelly, brown bread, elderflower cordial and sloe gin. Her recipe for chocolate brownies also took the gold, even though it was her sister who actually did the baking on that occasion. The sister still basks in her triumph but the Beloved knows that the glory is secretly hers. She is a very serious contender. Once or twice she has had a casual try at a few obscure categories like 'most tastefully wrapped gift' and 'healthy salad on a plate', but these are for dilettantes and this year she is returning to serious territory: tea brack, marmalade, shortbread and, because she now feels proprietary about the vegetables, beans (any variety, five on a plate).

In past competitions I have entered what I considered to be the best of my various chutneys. Every year I make a few varieties, depending on what there has been a glut of. I naturally make them to suit my own taste and think them all un-improvable, so it is always hard to decide which one is the most likely prizewinner. One year the Beloved and I, along with a chanteuse friend who was staying with us, had a tasting on the eve of the show. Chanteuse had brought a lot of Prosecco and she insisted we get our taste buds into shape by glugging substantial quantities. I set the table with some crackers, a little cheese and samples of my different batches, then we set to. Chanteuse felt we should cleanse our palates between mouthfuls using the Prosecco. Things got a little confused. We grew indecisive … and then divisive. I started blending chutneys in an attempt to find a compromise. A favourite combination emerged but it was felt that the pieces of green bean in it were just a little too big. I threw that particular mix into the food processor, blitzed it and jarred it. We were confident we had concocted a winner and we congratulated each other by finishing the Prosecco. The following day my jar of green sludge came nowhere.

This year, Chanteuse's summer visit has been postponed till after the show and I am not messing with anything. I am entering three different chutneys – all delicious – surely one of them will make it? My fantasy result would be to sweep the board and get first, second and third and then retire from public competition for life. Show entries have to be made by a particular date in order for the programme to be printed in time. In the past I have always called in to the house of the woman responsible for the home-produce section to deliver our entry forms and pay our entry fees in cash in person. This year, being up in the city so much, I missed the deadline. The Beloved rang me to say she had entered online. This seems somehow wrong. I am beginning to feel anxious.

The wound from my minor operation has finally healed but I have had a couple of other hiccups with my health – my right ear and left eye have been acting the maggot. I went slightly deaf in one and slightly blind in the other. Having once been severely sick, I tend not to let minor glitches worry me too much. They are irritating, though, and sometimes feel like the heralds of advancing decrepitude. The ear problem was easily explained and swiftly solved: a lump of wax, dislodged by my frequent swimming, had blocked a channel and affected my hearing – a visit to an audiologist with a syringe of warm water soon dealt with that. The eye was of slightly more concern: I had a vitreous haemorrhage, which is a little bleeding into the jelly bit inside the eye. I remember the precise moment it occurred – it looked like the spurt of ink from a fountain pen into a glass of water. I pretended it hadn't happened for a few days, then went to see my GP, who sent me to the emergency clinic in the eye hospital pronto. An ophthalmologist examined me with various complicated machines and reassured me that I wasn't about to go blind, but that I had 'traction on the inferior macular arcade'. The doctor said this was quite rare, which is gratifying – it is always nice to be thought interesting. It was going to need regular observation until it cleared up. As it happened, I had cameras trained on the retina and the eardrum in the same morning and was shown the pictures from both. It was intriguing and slightly disconcerting to see such detailed, magnified interior images of myself. The eye reminded me of a photograph of the sun: a serenely pale yellow disc with a network of orange capillaries, like flares on the solar surface. The hairy, darkly glistening recesses of my inner ear resembled nothing so much as a giant spider's arsehole.

In the dressing room that evening after the show, I recounted the results of my ocular and auricular examinations. Thus encouraged, the gathered venerable actors – the bearded one, the balding one and the one who is both – eagerly compared ailments. It seems I have much to look forward to. Thankfully, the conversation changed course when we joined the rest of the cast and crew in the bar prior to a company dinner. We had three birthdays to celebrate and we had booked a little Italian restaurant just along the boardwalk from the theatre. A family feel develops on many shows and gang outings are an enjoyable way to keep the boredom at bay over a long run. Backstage competitions are also common. I have known hotly contested table tennis, darts and poker tournaments, with heavy money changing hands in the knockout stages. Cheap tin trophies and prize-giving ceremonies are frequent. In one show, during which a fiercely fought and long-running quiz kept us sane, the leading team on any given night were allowed to wave their arms in triumph during the company dance at the curtain call. The choreographer only spotted this infringement on the last night, by which point it was too late for her to complain. In one interminable run of a wigs, swords and cloaks period piece, betting on anything and everything became the craze. Serious money changed hands over obscure issues. One actor wagered his house that a colleague would never, ever get a round of applause on a particular only-mildly-amusing line. Short of exposing himself, the actor concerned took suggestions from all round and tried every known trick to milk the line. Sadly, the audience never clapped and no one was evicted. Another wager between this pair was 'Who has the biggest head?' They meant by actual volume, not size of ego. Many bets were laid and, for a week or so, people would wander into the dressing room we shared and examine the heads concerned in studious silence, sometimes even feeling the respective craniums before risking their cash. We settled the issue on a Saturday between the matinee and evening performance. I was the judge.

In the wings, the crew had set up waterproof matting and two large, identical buckets of water, full to the brim and standing in deep-sided trays. The contestants emerged like boxers taking to the ring for a championship bout. They wore matching white vests with black shorts and each had their 'seconds' rubbing their shoulders and urging them on. I oversaw the attaching of strips of gaffer tape to both of the contestants' necks – precisely placed over the Adam's apple to serve as a water mark. Then they both immersed their heads up to the level of the tape whilst my assistants ensured there was no dunking below the line. The displaced water being equal to the volume of the head concerned, as Archimedes demonstrated, the tray with the most spillage decided the winner. It was a close thing, but my adjudication was clear, definite and could not be appealed. The fact that large drinks were bought for me had no bearing on the case.

Sometimes, to keep yourself and fellow cast members amused, it is hard to resist the temptation to adorn your performance by looking for extra gags. This is justified by telling yourself that you are enhancing the production. This play has been getting plenty of laughs at appropriate moments but I was convinced that we were missing one. I spoke to a colleague and suggested that he should glance in my direction on a particular line, we would both pause for a micro-second and I would give a reaction of great subtlety. This would, I assured him, release hilarity. He was willing and we tried it. In the silence that followed I imagined the sound of a wet sponge hitting the stage. Other 'enhancements' are more noticeable. I remember a big, swirling ensemble piece we did at this theatre some years ago: after an absence of a week or two, the director came to a performance, saw the large cast afterwards and asked us all to 'cut the improvements'.

Each audience is different and so each show is unique, but nonetheless they blur into one in the memory. As the run approached its end, the theatre arranged something which stands out. The play is haunted by the music of Chopin, and one evening after the performance, the pianist who made the recording gave a recital of the pieces from which the excerpts had been taken. Those that wished – audience, cast or crew – were invited to remain behind to hear it. It was a hot Saturday night and I was eager to get home but this sounded like a rare treat, so I stayed. I slipped in at the back as discreetly as I could. The house lights were up and I could see that most of the audience were still in their seats, and many members of the theatre's staff were leaning against the back row. I joined them. Just in front of me, a young couple had slightly shifted their position and she was now sitting languorously in his lap as they listened. The lighting on the set suffused a soft summer glow and the air, warmed by a full house and theatre lights, had a heavy, drowsy quality. The swirling notes of the first two pieces cascaded around the auditorium in a febrile flourish. It was the sort of music which must have prompted the remark that Chopin was best suited for the sick room – fevered arpeggios interspersed with brief, lucid high notes – lovely, voluptuous stuff. The final piece had a skeletal, wistful beauty. After the last, lingering chords had faded away there was a substantial, satisfied silence … Nobody wanted to break the spell with applause.

SUMMER MINESTRONE

Ingredients

2 garlic cloves, peeled and chopped
4 sticks celery, chopped
3 small red onions, peeled and chopped
4 tbsp olive oil
1kg broad beans, shelled
1kg thin, tender asparagus cut in 1cm pieces
500g green beans, trimmed and chopped
500g fresh peas, shelled
1 litre homemade chicken stock
salt and pepper
bunch of basil, marjoram and mint, finely chopped
200ml cream
150g Parmesan, freshly grated
120ml pesto

Method

In a heavy saucepan, fry the garlic, celery and onions gently in the olive oil until soft. Divide all the other vegetables between two bowls. The vegetable quantities above are only a guideline. Chances are you might not have that much asparagus to hand so bump up the quantities of the other vegetables. Add one bowl to the onion mixture and cook, stirring, for a further 10 minutes. Season to taste. Cover with the chicken stock and bring to the boil. Simmer for 30 minutes.

Add the other bowl of vegetables and cook for a further 5 minutes. Remove from the heat and add the herbs, cream, Parmesan and half the pesto. Stir and serve with a teaspoon of pesto on top. Will feed about 8 people. Serve with crusty bread for mopping the bottom of the bowl.

Falling Sick

Our two family festivals in late July are birthdays – the Beloved's and then Younger Boy's to end the month. The former fell midweek this year. I drove home after the show in order to give her breakfast in bed (which she actually enjoys) and to orchestrate her birthday tributes. Another summer's morning dawned. The sun was bouncing off the white walls of the house when the boys and I went out early to pick wildflowers: yellow poppies from the bank; orange montbretia from the grass verge; and the white trumpets of bindweed from the hedge. Then we made cards and breakfast, retrieved the gifts from their various hiding places and headed up the stairs, boys first. Their excited grins of anticipation as they burst through the door are pretty hard to top as a birthday treat, but she played her part well, pretending surprise and making all the right noises as she received cards, flowers, poached egg on toast and presents. Older Boy gave her a tasteful multicoloured bracelet with the letters of her name stamped on lurid plastic beads. The younger one gave her a giant lollipop with 'Number One Mum' emblazoned on it. I had taken them shopping the previous weekend and he had pointed out that if it proved too big for her he could always help her eat it. I gave her some scent from a funky parfumier she likes. The fragrance I chose used to be her favourite, but not any more – apparently there are new aromas in the range. She was delighted that I had got it slightly wrong so she was therefore able to look forward to exchanging my present for something else. So my choice of gift had not been perfect, but was almost so, which was actually … perfect. I don't really understand.

After breakfast I made a birthday cake – just a simple sponge: easy to whip up and delicious when freshly baked. Apart from the Christmas fruitcake, it is the only cake in my repertoire, but it is a perpetually useful, all-purpose festive classic. I rang the Farmer Friends and told them there would be coffee and cake at eleven. He was away but she would be there. Polytunnel Pal was already coming over with a gift of brassica seedlings – cabbage, sprouts and broccoli – but I thought we needed more of a gathering. The village choir mistress was taking an exerting walk up the hill so I told her to drop in on her way back. Two joggers passed. One was my chairman's wife and the other was one of the Beloved's sisters-in-arms from the battle to build the playground. When I suggested cake they made noises about replacing calories that they were attempting to jog away. I told them it was a very light homemade strawberry-and-cream sponge and they agreed – just to wish her a happy birthday. We had an impromptu party under the big cherry tree. The cake was quickly demolished and Younger Boy got more than his share of the lollipop. Later that afternoon the Beloved hinted that I might enjoy planting out Polytunnel Pal's present, as I had missed so many of the tasks in the vegetable garden this summer. I suggested that she might appreciate the gift more if she did that herself. Besides, I had to head back to the city for the evening show. It was another lovely drive. The predominant colour in the hedgerows was the purple of the salvia, or wild sage; the yellow grass in the verges was flecked with green clumps of thistle and dock; and the golden cornfields stretched for miles. The Beloved rang me at the theatre that night to thank me for her birthday treats and to tell me that she had finally put the netting over the brassicas.

I have only failed to mark her birthday once, but I suppose I had a tolerable excuse. Fourteen years ago, I spent two long summer months in hospital fighting to recover from a grave illness that apparently nearly killed me. Twice.

I had been working regularly at a theatre in the heart of the city in overlapping productions of Dickens, Wilde and Shakespeare. We were rehearsing a Shaw play and I hadn't been feeling well – upset stomach, abdominal cramps and occasionally urgent dashes to the loo. I ascribed it to late nights, irregular eating and too much roistering. I was tired from the long stretch of work and thought it was just my slightly chaotic and admittedly unhealthy lifestyle taking its toll. I told myself that I was just a little run-down. Once this latest play was up and running I intended to go easy and adjust my diet. During previews I began to get waves of nausea and was also finding it hard to eat, which was – well, astonishing. The theatre doctor was called. He gave me something for the nausea but said I had something serious going on and should get it checked. I had a play to do and nodded and said I'd get onto it. The theatre's director intervened and had me swiftly seen by a top gastroenterologist who hospitalised me the morning after we opened.

I rang the theatre from the hospital with the news. An hour passed and they rang back saying that they had arranged for another actor to take over my role, but that he would not be available for the next two scheduled performances. I was feeling pretty groggy and grey around the gills at the time, but I remember two distinct feelings. One was a sneaking self-regard at the high calibre of my replacement and the other was a determination to do the two shows in the meantime. I told the consultant my plan. He wasn't happy. He explained that a significant length of my gut was, I forget the medical term – banjaxed, was it? Anyhow, he wanted to get me straight onto intravenous steroids to reduce the inflammation and gain time for more considered treatment. There was no doubt in my mind that I was doing the shows – to let my fellow cast members down felt impossible to me. Revered consultants are powerful people but this one knew he couldn't hold me against my will, so he begrudgingly acquiesced to my suggestion that I merely be detached from the drips for a few hours whilst I was driven to and from the theatre to give my performance; otherwise I was under his orders. The Beloved felt I was a fuckwit but she knew I was obstinate and, having seen me settled into a bed and hooked up, she went off to borrow her sister's car. That evening, I called over the young doctor who was on duty and asked him to detach the drips from my arm as I had somewhere I needed to be. His eyes danced in disbelief. He started to say, 'But you don't understand –' I cut him off to explain that actually he was the one who didn't understand. Poor lad – he had his own, extremely laudable professional duty of care to enact and he couldn't grasp that there could be a supererogatory imperative. I told him that I had squared this with Revered Consultant and hobbled out of the ward.

I gave my best performance in the role that night. In fact, I got a small round of applause at the end of one of my scenes and then limped back to the little camp bed that had been set up for me near the stage left entrance. Whilst on-stage I forgot my symptoms – I was aware only of the dialogue and the characters and the movement of the action – 'Doctor Theatre' we call it. Offstage I lay on the bed, groaning slightly from the insistent ache in my belly. The Beloved was right – I was an utter idiot. Still, at the time I felt it had to be done. When she drove me back to the hospital again after the next night's performance, I had no option but to surrender to sickness and the medical machine's efforts to save, heal and cure me. They achieved the first two but not the last. As it turned out, I had contracted an illness from which I will continue to suffer till the day I die. But the worst is hopefully behind me. Revered Consultant did his utmost to avert the surgeon's scalpel. He switched me to supercharged steroids and examined daily X-rays in the hope of soothing and repairing my massive intestinal inflammation, and indeed I started to feel better, but from the increasingly frequent visits of the surgical team as the days passed, I began to realise that the drugs weren't working. After a fortnight they operated and took out a substantial length of my digestive tract. The surgeon afterwards described it as having the consistency of wet tissue paper. The procedure took twice the anticipated time. When I returned to a vague consciousness in the Intensive Care Unit, I was minus much of my intestines. In their place was an ileostomy.

I tried to ignore this unwelcome, upsetting addition and was largely successful, partly because I was in a morphine haze and partly because all the other paraphernalia camouflaged it. I felt slightly detached from my body and observed the various tubes coming in and out of it with a bemused objectivity. This was no doubt because of a subconscious psychological defence mechanism – my body had taken a severe battering and my mind was doing its best to dissociate itself from all this unpleasantness – but also, I have to say, the drugs were very good. My morphine was self-administered. I had a little hand pump with a button – when I pressed it a shot was delivered intravenously. It was very precisely calibrated by a little attendant machine, which regulated the flow to prevent inadvertent, or indeed deliberate, over-indulgence. For the most part, I could get just the amount required to keep the pain at bay. But every once in a while maybe a tad more than necessary trickled through and everything felt really, really fine.

However, things got worse before I got better. The huge amounts of powerful steroids I had been given before the operation severely compromised my body's ability to heal itself. The tissue along my scar was failing to knit properly. Peritonitis had set in – a grave concern with anyone suffering from underlying illnesses, and in my case, as I later discovered, very nearly fatal. Another operation was required. Ten days after my first visit I was back in theatre to have, put simply, my innards pulled out, scrubbed and put back in again. This time I remember coming around in recovery with my mum and the Beloved on either side of me, each holding a hand as I whimpered under waves of exquisite pain. I had become inured to the morphine and, despite the horse's dose I had been given, I was feeling an agony I would wish on no one, not even that casting director. In due course, I was given a different powerful palliative and I relaxed once more into a narcotic limbo.

Back on the ward, I again observed the tubes with disinterested curiosity. Extra ones had been inserted whilst I was under the anaesthetic. Because of the assault on the remainder of my intestines, they had gone into shock and had temporarily stopped working. I would therefore be unable to eat and was going to be fed intravenously through a central line – which is a long, fine tube inserted deep into the subclavian vein. Digestive juices would nevertheless continue to be produced and they would need to be extracted, so a nasogastric tube had also been fitted – up the nose and down the throat: not pleasant. There

was a catheter for urine; the tube from the stoma carrying away faecal waste; and delivery systems for blood, oxygen, morphine and fluids. I can't really recall if they were ever all working at exactly the same time but there was a possible maximum of eight different substances being siphoned in and out of me.

I was bedbound for weeks and withdrew deep inside myself. I was at the end of the ward and through the large windows I watched the summer sun's daily journey over the Dublin Mountains. I drifted in and out of wakefulness. The days were warm and cloying; the nights were fractured and cool. Light and dark interpenetrated one another. There was the constant opulent aroma of the flowers that just kept coming, the steady hum of machinery and the dedicated industry of the nurses. I was superbly cared for.

When the fluids in my stomach – the bile – accumulated, the only way out was up, and once or twice I vomited the vile green liquid everywhere. I learnt to press the bell at the first hint of discomfort and someone instantly appeared to relieve the pressure by drawing off the stuff through the nasogastric tube. Occasionally other waste tubes became compromised and I would awake from a fitful doze to find myself in a pool of my own excrement. I felt squalid and pathetic. The nurses cleaned me up with cheerful chat and smiles. They washed me daily; they changed my sweaty sheets; they fixed a fan above the bed and fiddled with it till it gave the precise level of gentle cooling breeze I needed. They literally tended to my every need. They nursed me. I know it's in the job description and it is what they are (inadequately) paid for, but I will be forever grateful for the care I received from the staff of Crampton Ward, Tallaght Hospital in July and August 2000.

In my mind's eye, that summer is an amorphous, flickering blur of image and mood, yet at the same time it has a palpable presence in my memory.

I remember the afternoon I sat out of bed for a while.

It was a hot day but I started to feel cold. My shivering became uncontrollable shaking. The nurses came running, then doctors too. I was lifted bodily into bed. Fond Aunt appeared – she had come for a visit. I glimpsed a stricken look on her face as the curtain was whisked around me. Something was seriously up. I felt I was freezing and I shook like I was in a fit. I couldn't understand why they were placing icepacks around me. My temperature had shot through the roof as a result of an infection caused by the central line deep in my chest. The fevered shaking, or rigors, was brought under control, I was given some major antibiotics and also some Valium and I floated to somewhere serene and smooth. Apparently that was the second time I nearly made my exit.

I remember the first day I managed to wash myself without assistance.

I had become used to, and not embarrassed by, my mid-morning bed bath, and on this particular day the usual time for it had come and gone. I looked down the ward to the nurses' station. A few of them were hanging around, avoiding my eye and pretending to look busy. I realised they were waiting to see if I would show initiative and walk to the bathroom unaided to perform my ablutions alone. I decided to try. I picked up my wash-bag and set off slowly down the ward, pulling my mobile intravenous drip behind me. The nurses fell silent and surreptitiously watched as I shuffled past. In the bathroom I saw myself in the mirror for the first time in well over a month. I was appalled. I looked skeletal. My legs and arms were sticks, my buttocks were loose flaps of skin and my face consisted of two huge startled eyes and a rack of teeth. The shock far outweighed the pleasure of noticing that I now had great cheekbones. I sponged myself, cleaned the scary teeth and made my slow way back to bed, proud of my achievement but chastened by what I had seen.

I remember my mother sitting reading by my bedside day after day. I only discovered recently that Fond Aunt had visited daily, bringing food for her so that she didn't have to trudge the long walk to the hospital canteen.

And I remember the Beloved giving me a foot massage.

I am still touched by the love and care my mother showed for me during those months, but I understand it. Now that I am a parent I can completely identify with that compelling necessity to stay by your child if they are suffering. What I am still in awe of is the Beloved's intense, yet calm, perpetual presence. My mother tended to stay for the afternoon and she took over in the evenings. One night when I was beginning to feel more conscious of what was happening to me, I unburdened myself to her. I think at that stage I had an extra tube in me, draining surplus fluid from my wound. I could hardly move because of all the medical plumbing wrapped around me, but knew I didn't have the energy to move much anyway. I couldn't eat or drink anything and, despite the intravenous fluids and an endless supply of damp mouth swabs, I felt an unceasing raging thirst. I was trying to regain a little lucidity and was therefore avoiding over-reliance on the dull opiates. As a result, I was slightly more clear headed but I was also in constant discomfort and occasional fierce pain. I hadn't had a second's ease, let alone pleasure, since an age ago in what seemed like another, past life. It was all ghastly. She didn't need to be told – she had watched all this happen – but she listened. The next evening she came in with scented oils. She unrolled the constricting knee-length stockings I had to wear to prevent thrombosis; then she anointed my shrunken feet and gently rubbed in the oil. I drifted on deep, even breaths and slowly unwound.

I remembered that I had forgotten her birthday and apologised. She smiled and told me what she had done. It had fallen on the day after my first operation. Her mother had taken her for dinner, not to celebrate – she was in no mood for that – but for company. In response to her mother's enquiry, she had drawn a little sketch of what the surgeons had been doing to me. It was a rough approximation of the particular stretch of entrails I had

had removed. At this point her sister joined them. The Beloved showed her the drawing and asked her what she thought it was. 'I dunno!' was the reply. 'Earrings?'

Finally, I remember a fragment.

It was in the small, slow hours of the morning. The night nurses were quietly dealing with notes and charts in a pool of light at the other end of the darkened ward. The crisp bedsheet was warm and dry under my body, and the air above me shifted gently in the soft, cool breath of the electric fan. I was aware of the grip of my surgical stockings, the tug of the adhesive dressing on my chest and belly hair, the hospital gown loosely draped around my shoulders, the light touch of the plastic tubes going in and out of my arms and neck, the pain of my wound, which I acknowledged and accepted ... I was conscious of many precise bodily sensations, whilst at the same time my mind ranged in long, stilt-walking strides.

I had been pondering what I had only recently discovered: that I had nearly died on the operating table. I was contemplating this quite calmly – I have no doubt the drugs aided this equanimity – and I was wondering if I would have been satisfied with my life if that had been it. I decided that I could have ticked a lot of boxes: I had travelled, thought and felt. I had chosen a profession I adored and once or twice done work I was proud of. I had loved some remarkable women. I had made good and interesting friends. And lastly, integrally, I had met the Beloved. The one obvious absence I was aware of was that I had not become a father. But that may yet be in my future, I thought, given that I still had one. I became aware of my steady breathing and the clean, salty nip of the oxygen from the tubes beneath my nose.

Then I had a glimpse.

I sensed the universe: its vastness and its simplicity; its chaos and its order; its absurdity and its perfection. And I laughed. It was so brilliant, and obvious, and hidden under my nose all this time. I grasped its essence. For a full, intuitive instant, I understood. It was the best, most joyous joke ever. I laughed. And then it was gone. I haven't a notion now of what that brief phenomenon was – not a taste nor a hint of its detail remains. Fractured, disparate thoughts and feelings coalesced for a moment of ideal clarity and then fled, leaving merely a sense of an outline. I cannot remember what I briefly knew. But I will never forget that laugh.

That stay in hospital now seems somehow unrelated to the unfolding calendar of my life. It feels 'other', separate and out of time. It is only now, as I write, that I realise it too was once situated in the cycle of a year ... this time of year ... now. But that was then. It has never since had an association with the Beloved's birthday: it is behind me, gone, yet it will always remain present in the past.

I was admitted to hospital a fortnight before the Beloved's birthday. When I left in her care, two months later and four stone lighter, I knew I would acknowledge the day for the rest of my life. We were in a steadily growing relationship when I fell ill. By the time I was beginning to recover, I intended never to take my eyes off her. Voluntary ties, growing stronger with the passing years, have kept us together. We are unmarried, and will possibly remain so, but I can imagine nothing in any civil or religious ceremony that could make our connection more profound.

Having said that, I was raised in a Christian dispensation, and a Catholic one at that, and the patterns of one's thoughts are inevitably culturally informed. In the moments of the deepest crisis of my illness, when the pain and struggle seemed unrelenting and inescapable, I 'offered it up'. Not to Jesus, but to some vague notion of karmic balance in a just cosmos. I made a deal. I agreed to suffer this if she could be spared the pains of childbirth should we ever decide to have any children. It didn't work for Older Boy's birth, but perhaps there was a deferred reaction to my bargain because the second boy arrived in a fashion much closer to how the Beloved hoped his birth would be.

We were in the flat in town the night before he came. I erected the birthing pool and went out to get a takeaway curry. We were aware that spicy food supposedly encouraged labour, but actually it was just what we fancied. We ate, then she retired to bed and I went out for a stroll. I crossed the river, walked along the boardwalk and had a glass of wine and a cigar at an Italian cafe. The water rippled sluggishly, the cyclists, the traffic and the pedestrians passed, I sat and waited. I scribbled on the back of an envelope – snatches, impressions and half-formed phrases: an attempt at a poem to pin down the moment and my mood. I have since mislaid it and it was most likely twaddle, but the act of writing something is in itself a record; even if what is actually written is lost, the formed words stay in the mind. I set it down and it roughly remains in the tables of my memory. The slow passage of the river at my elbow, the people and the vehicles swirling around – all these wanderings seemed frivolous compared to the journey that was nearing completion on the other side of the bridge: the true trajectory of my second son. The Beloved and I had made love, I remember the moment, on the eve of Samhain in that flat across the water – I could see it from where I sat – and the boy was about to be born on the eve of Lughnasa, a few feet from where he had been conceived.

The following morning, the same mighty midwife who had

overseen the first birth called in. She made her usual thorough examination and said that the baby could arrive any time soon, but not right this minute. We decided to head back down to the country. We collected Older Boy from his grandmother's house near the larger neighbouring village and took him home. He and I rooted about in the kitchen garden. The Beloved mooched around in the kitchen. She made a puttanesca sauce for pasta and then came outside to join us in the late-afternoon sunshine. We were picking peas when she suddenly went, 'Oh!' and made a startled expression. It was time to go. I rang Mighty Midwife to alert her. Then I grabbed the pasta sauce, locked the house and started the car; we all piled in and headed smartly off. I stopped for an instant to return Older Boy to his grandmother and drove with a concentrated urgency back to the city. The Beloved started to groan intermittently, and then regularly. The groans became fierce grunts, the grunts got louder and then the bellows began. I remembered those. At one point, she roared, 'This car is really uncomfortable!' I thought to myself, 'Yes, my dearest, that's what's discomfiting you – it's the car.' Out loud, I agreed. The car was indeed old and poorly designed. 'Be quiet!' she barked. I drove with as much speed as was consistent with safety and, above all, smoothness. I wondered if the battery on my mobile phone had enough power to sustain a long conversation with Mighty Midwife, as I acknowledged the possibility of her having to talk me through delivering the baby on the roadside.

As we approached the city, I rang the Beloved's sister and told her to be ready at a particular street corner in so many minutes – I would need her. She was there at the appointed time and place. I pulled over, she hopped in the back and we sped off. I started to give her instructions, speaking forcefully but calmly to soothe

the obvious bewilderment she felt at the astonishing noises her sister was making. She was to park the car after we had jumped out at the flat, and then go and get solid and liquid sustenance – if the labour was to be long, the Beloved would need it. The sister looked at me uncertainly in the rearview mirror, still a little shaken by the sounds from the passenger seat.

'Get Lucozade and yoghurt,' I said.

'Plain low-fat yoghurt?' asked the sister.

'No! Full fat!' roared the Beloved. And then an afterthought: 'Strawberry!' she shouted.

There was a brief silence. The sister started to make conversation.

'Stop wittering!' yelled the Beloved.

I could see the sister was a little ill at ease. I started to make idle chat to draw fire. It came.

'You can shut up too!'

Inside the flat, the Beloved hit the floor and I started to fill the birthing pool. Mighty Midwife arrived minutes later, sized up the situation and told me not to bother – the baby would be here before it was full. She attended to her task and I brought up her heavier gear from her car. It included an oxygen bottle which was never used. We got to the flat just after seven in the evening and the boy arrived before nine. There was another person in the room and no one had come through the door. She had given birth where she wanted to and how she wanted to. No drugs and no intercession.

I took a picture of him in his mother's arms when he was maybe a minute old. I see it now. He looks like every newborn baby ever looked. She looks radiant. Her hair is lustrous, her eyes are flashing and she is laughing. She was also ravenous. I cooked some spaghetti and threw it into three bowls with the puttanesca sauce. Mighty Midwife and I ate ours gratefully. The Beloved sat on the sofa with the boy under her arm whilst she ingested forkfuls with appetite and relish.

Seven years ago.

This year my mum came to visit for his birthday. I met her off an early flight from England and took the inland route south. We left the environs of the city and drove through endless yellow cornfields demarcated by the green strips of hedgerows. The small car felt cramped – not by mum nor her small bag and large presents, but by the giant helium-filled, golden balloon in the shape of the number seven which I had stuffed in the back. We let it trail out the window of the car as we pulled up to the house, horn blaring.

His party was smaller than last year's but the giant, golden seven was taller than him and he thought it was splendid. The Beloved's mother joined us, so he had two grandmothers doting on him and the best balloon ever. I lit the barbecue and we had burgers followed by chocolate roulade as he had requested. He enjoyed his day … until the balloon floated away over the hills. He and his brother had discovered that it made an ideal sparring partner for Taekwondo: it bounced back but never hit back. Their laughter from the bottom of the field was suddenly pierced by a wail. Older Boy had apparently given it an extra potent roundhouse kick which detached it from its weight, and it swiftly made its escape. It looke'd quite pretty as it bobbed gently up into the blue sky and white clouds, twinkling in the sunlight until it dwindled and faded from sight. The boys were distraught: the younger one howled over losing his balloon and the older one blamed himself bitterly for ruining his brother's day. I assured the latter that he hadn't and the former that I would buy another balloon and would bring it with me next time I came home. They both cheered up. The birthday boy decided that now he was seven, he should put away childish things, and he gave me an old cuddly toy mouse to give to Fond Aunt so that she could pass it on to his younger cousin, her grandson. I put it in the car, along with samples of my chutneys for my dressing-room companions to consider, and made the return journey to town for the show.

The Venerables tasted the chutneys with due attention and discernment. They asked searching questions about the blend of ingredients and the balance of spices. I referred to the apple and courgette mix as my 'flagship' chutney. They named the other two 'Admiral's Choice' and 'Bosun's Delight', and they assured me that I would torpedo the competition and that other entries would sink without trace. The nautical metaphor was extended until I threatened to flog them. They piped down.

I made one last round trip to the country from the town before the play closed. I brought Fond Aunt down for the day so that she and my mum could catch up. I also brought another golden seven. The boys were delighted to see the balloon and the women were delighted to see each other. I left them chatting on the bench at the front of the house, looking across to Sliabh Buí, whilst I went up to the kitchen garden to pick something for lunch. I wandered through the profuse greenery at my leisure and savoured the return of the season of abundance. The corn was as high as a baby elephant's eye; the peas were cascading from their canes; the beans were pendulous and plentiful; the proud potato plants were in purple and white blossom; the cucumber tendrils snaked across the path; leeks, turnips, carrots, beets and parsnips were in variegated lines of green leaf; and there was an explosion of lettuce. But the brassicas looked buggered. Under the netting, a horde of cabbage white butterflies whirled and fluttered around totally defoliated stalks. The eggs whence they had emerged had obviously been laid before the netting was placed. The caterpillars, thus protected from marauding birds, must have hatched in peace and found themselves trapped amongst limitless food. What else could they do but gorge themselves to happy oblivion? I decided not to say anything to the Beloved and sat for a while in dappled shadow under the silver birch to stifle bad thoughts. Behind me loomed the fuchsia hedge, fat with bumblebees. In front of me was a prehistoric swamp of rhubarb, and beside that rose the first feathery wisps of the asparagus which we hope to harvest two years from now. Above me crows cawed lazily, swallows swooped in the heavy, still air, and over the warped roof of the nearly collapsed hen house, I could once again see the surrounding hills display that familiar mosaic of green and gold. I picked some fine looking beans and brought them into the kitchen. The Beloved looked up, knitted her brows and bit her tongue.

'What?' I said.

I had harvested the prize specimens she had been saving for the county show. I mumbled an apology and went outside. It was definitely not the time to mention the cabbages. Mum and Fond Aunt were podding peas and the jubilant yells of the boys suggested they were giving the new balloon a kicking. Then there was another wail and another golden seven winked and glittered as it floated up and away.

On the drive back to town for the last shows of the run, I realised it was the first of August, Lá Lughnasa – the festival of the pagan god Lugh. I cast my mind back to the play I did in London two decades ago now, unfolding as it did during this time of year. I remembered the set: a country cottage surrounded by a swathe of ripe corn. Just like ours. I thought of the play I was just about to leave behind, by the same writer, also set at this season and steeped in the music of Chopin. The symmetry pleased me and then the coincidence deepened. As I passed through such a field of gold, redolent of the high point of my career in England, I turned on the radio and heard the elegiac notes of one of the Chopin pieces I had been listening to all this summer on the stage I love best in Ireland. I smiled as I drove. The set, the music and the plays were artifice, but the corn around me was not. The imagined August was ending as the actual one was about to begin. The written season is giving way to the real and the harvest has come again.

LUGHNASA REDUX

The Harvest Returns

Judging Chutney

Woken abruptly by birdsong on the August bank holiday Monday, I was in a strange, unsettled state of mind. I felt optimistic yet anxious. I opened the bedroom shutters. Bright sunlight flooded the room and the kitchen garden was a cornucopia. Then the birch tree started to sway, the sky darkened and an armada of grey clouds appeared over the top of the hill. Our local microclimate is mercurial enough to confound most meteorologists, and I am suggestible enough to be emotionally influenced by changes in the weather, but my variable mood was because of something more substantial and profoundly uncertain. It was the day of the county show. The Beloved and I had made our entries. In a few hours the judges would decide. Was triumph imminent or did failure beckon?

I have dim childhood recollections of holidays in the country and being dragged around a sodden field on fair day in Ballybugger to look at a goat with three horns and a limp bull. And for most of my adult life, the notion of a day at a country fete would have made me shudder. But I have changed. 'The Show' is now a red-letter day in our calendar. The first time we attended, we'd been here less than a year but had grasped the nature of the microclimate and were suitably attired in shorts, wellies, sunglasses, rain-hats and mittens. We had a great day. There were majestic bulls and mighty cows. There were oily-fleeced sheep and powerful rams. There were goats galore – none with three horns, sadly, but many with the biggest testicles you ever saw (the Beloved just does not understand this fascination). The ranks of polished, gleaming agricultural machinery glinted in the intermittent sun. There were vintage cars and fairground rides and candyfloss and beer. There were talent shows and dog shows and show-jumping and geese. Delicious gourmet snacks were available alongside burgers, chips and pizza. The stalls and marquees sold everything from tweed coats and windmills to teapots and jewellery. We ambled about gently for an hour or two taking it all in. We took it handy, as they say around here – had a cold drink, chatted to anyone we knew and some we didn't, and nodded our heads to the beats of the boys in the band who were performing from the back of a lorry. Their audience was sparse and perched on straw bales, but they played like rock gods transfigured by stadium lights. Later I passed them on the way to the loo. They were still pumping it out, although a shower had started and most of the small crowd had scattered to the shelter of the beer tents. As I walked by I heard the singer shout to the unoccupied bales, 'Come on, everybody, all together now!' There was the gig happening in his mind and there was the gig happening in the field. He knew which one he preferred and I wanted to cheer him. We wandered into the home-produce tent to admire the jams, cakes, pickles, loaves and vegetables. We both thought it pleasing that country skills were alive and well. Then it occurred to us that we might enter something ourselves the following year. Then we were screwed.

We began our competitive career by submitting our son for consideration. There was a bonny-baby competition, so we entered Older Boy. Just for a laugh, of course – we were not taking it at all seriously. This is merely an amusing diversion, I thought, as I hissed to the Beloved to hold the boy facing out, for god's sake: the judges were coming back for another look! He came second and, oh, the relief! I might have demanded a steward's enquiry if he hadn't been placed at all. The next year we entered chutney, soda bread, two types of jam and a photo of the boy running along the sand. We won nothing. Not a rosette between us. We were stunned, staggered and outraged. Then we calmed down and laid our plans for the future. We noticed that certain names completely dominated the baking and jam-making categories. They seemed impregnable. We needed a new category. We came up with 'Hedgerow Produce'. This would cover elderflower cordial, crab-apple jelly, sloe gin, blackberry jam and anything else, basically, that one can forage for. We figured we could make a strong showing in all of these items. After assiduously flattering the local librarian, who had a soft spot for our son and, vitally, was on the committee for such things, we had the category established. Since then, we have regularly been amongst the rosettes. There have been occasional disappointments – we are never presumptuous – but this year we were quietly confident and I especially hoped for glory. The fact that it is the 'county' show means that if I were to win first prize I could justifiably claim to make the best chutney in Wicklow, and that continues to be the ultimate goal.

But there is a more local aim. The Beloved and I have an unspoken but long-standing rivalry with two other couples from the smaller neighbouring village. We are all newcomers to the county, have children of similar ages and have enjoyed each other's hospitality over many convivial years. We often enter the same categories and on show day we usually meet up after the adjudication to have a drink and some friendly banter about who won what. The atmosphere is one of amused irony, casual indifference and tacit, fervent contention. The Beloved frequently goes head to head with one of them in soda bread, banana bread and shortbread. Honours are roughly even. I regularly take on another in chutney and she always beats me. She never crows about it, but whenever I call in to her house for coffee she invariably seats me where I can get a good look at her accumulated rosettes over her shoulder. She enters scores of categories, to better ensure she ends up among the prize winners, and has been known to trawl the Internet at midnight, searching for the best way to tie up onions for exhibition. She takes it far too seriously. This year I got an anguished text from her. Sent at dawn on the morning of the show, it howled about a disaster with her jam. I laughed at her absurdity and then re-jarred my three chutneys to make them more alluring.

It was all a waste of time.

My chutney came nowhere.

Flagship … Admiral's Choice … Bosun's Delight … they all capsized.

I texted the Venerables – they were astonished and suggested keelhauling the judges. It got worse. The Beloved's marmalade, her shortbread, her tea brack, her beans – they all failed to be placed. It became unbelievable: none of our friends had succeeded either – not a single rosette between us. Our usual post-judgement drinks tasted sour and our banter became bitter and paranoid. Maybe the judges had tired of us arrivistes and our ambiguous attitude: interpreting our light irony for dark sarcasm and our sense of fun for a lack of seriousness. Perhaps they had mounted a vendetta to put us in our places. I tried to rise above these black thoughts but it was hard. As Male Farmer Friend put it later, when I told him of our debacle, 'Life seems fine then it bites you in the arse.' We wandered through the rest of the show with a less purposeful step. The boys enjoyed themselves despite their parents' disappointment. They sampled a bouncy castle and penalty football; ate candy floss, hot dogs and ice-cream; and blew a fiver each on wooden samurai swords. Later on we bumped into a stuntman friend who had entered his dog and his daughter as a team in the fancy-dress competition. They wore matching flamenco frocks of purple and silver. Our mood lightened a little and the day was gradually salvaged. Nonetheless, it was difficult to forget that my three best chutneys ever had been rejected. It was dismal. My faith in the judicial process has been undermined, perhaps irrevocably. I might walk away from the hedgerow-produce tent for good.

Once the show was over and the fair fields cleared of all the detritus, the county returned to the business of the harvest. All the agricultural behemoths – the combines, the tractors, the trailers – have taken to the roads again. The patterns in the fields are changing daily. Just before cutting, the corn extends evenly like the sea. Expanses of wheat and barley ripple tranquilly in mild breezes. After cutting, threshing and the separating of the grain, only the discarded stalks remain: they lie in long uniform lines and the fields look combed and braided. Finally, after a few days drying in the sun, the straw is baled and the familiar rows of huge golden cylinders reappear on the hillside. This process unfolds over weeks and in waves. People follow their various distinct schedules as their particular fields ripen. The four of us strolled out one warm afternoon. Male Farmer Friend was beginning to bale. We watched his progress along the bottom of the valley as we walked up the road. At the top edge of his land, a combine harvester crested the hill – a trundling, dark mass outlined against the sky. We recognised the neighbouring farmer by the silhouette of his cap and pipe. He lifted a slow arm in greeting. Scattered crows whirled in the sky and the road disappeared into the swathes of uncut corn like a vivified Van Gogh painting.

At home the hearth is filling once again. The stockman coat has been retired; it has declined a long way from the pristine elegance of its London life and its cowherding days are done. Now it hangs in the garage: crumpled, stained and smelling dubiously. But it is a proud and happy coat. It has fulfilled its proper purpose with grace and honour. The inglenook of the unused kitchen fireplace is stacked with jars of pickled shallots, plum jam and green tomato chutney. The freezer is stuffed with peas, beans and courgette soup. We pick salad leaves and dig potatoes when needed, the sweet corn is nearly ready and the early apples are crisp and sweet. Annoyingly, we missed the entire crop of cherries – birds ate the lot before they ripened. The trees may have been stripped of fruit, but the flocks still congregate. Each day is underscored by the

chirruping flight of swallows and the cawing army of crows. One morning the tall silver birch disappeared under a colonising horde of house martins. They fluttered and gathered before departing like a swarm of bats in a vampire movie.

Lughnasa's abundance can linger for weeks, but my enjoyment of the season is going to be cut short. The long, lazy August I had hoped for will have to be distilled into one swift week because I am returning to the city to begin rehearsals on another play. I am pleased to be working again so soon, and excited by a new and very different part from the last, but again I will be gone before I am ready to leave. I will not be indulging in slow afternoons of surfeit or warm evenings watching the barbecue smoulder.

My time at home will be brief – almost momentary. I must resist the urge to snatch at it, to grab at transience and thereby crush it. I have accepted the job – I must relax and accept the consequences, then I can fix these passing moments in my mind and make a mental picture to contemplate when I am missing home. I was standing in the sunshine at the high end of our field, taking in the laundry. I un-pegged a pair of Younger Boy's surprisingly small pants, then paused and smiled. No one was in earshot, so I allowed myself to go 'Aah.' I looked around. At the table under the big cherry tree, shielded from the sun by a giant umbrella, the boys, wearing only shorts, were quietly absorbed in building Lego. My mum was on the bench at the front of the house, shelling broad beans for dinner. A cloud of butterflies – peacocks, painted ladies and tortoiseshells – swirled and settled on the buddleia. I plucked a tea towel from the line and snapped it before folding. Bone dry, it made a crisp, pleasing whip crack. I looked across the valley to Sliabh Buí. The Beloved emerged from the house with a tray carrying a pot of fresh coffee, some elderflower cordial and a plate of award-worthy shortbread.

And I thought, that'll do.

RUNNER BEAN CHUTNEY

Ingredients

4 medium onions, finely chopped
4 hot chillies, very finely chopped
seeds of 30 cardamom pods
2 tsp fenugreek seeds
2 tsp cumin seeds
salt and pepper
1.5kg runner beans, thinly sliced
500g demerara sugar
450ml cider vinegar
neutral oil, like sunflower

Method

Fry the onions over a low heat until soft. Add the chillies and spices along with the salt and pepper. Fry for a few minutes to let the spices release their fragrance. Add the sliced beans, sugar and vinegar. Cook over a low heat until the beans retain a bit of bite but are not too raw. This could take an hour or more.

A top tip I was once given by a seasoned chutney maker is: 'Never embark on chutney making with a deadline in mind. It will be done when it is done.'

When it is done, put it in sterilised jars while still warm and seal. Don't forget to label the jars.

EPILOGUE

When I was a small boy, missing my father, I tried to fix many moments in my mind. The distance was so huge and letters so rare that my need for contact took refuge in fantasy. I imagined I had telepathic powers and that if I used these powers sparingly and precisely, I could send my dad a picture – mind to mind and in real time – of whatever activity or situation it was that I particularly wanted to share, good or bad, for approbation or reassurance.

I evolved a ritual that had to be followed minutely if this transmission were to work. I closed my eyes, concentrated hard on clearing my mind and inwardly intoned these words: 'I wish Dad could see from my eyes and knew he was seeing from my eyes, from … now!' Then I would open my eyes and he would see what I saw. It was important that I added the qualification 'from my eyes', otherwise he might be perplexed by these unattributed images popping up anonymously before his gaze. It was also vital that I said 'now' and opened my eyes at absolutely the same instant, otherwise the message wouldn't be sent. It didn't matter that he obviously lacked the power to communicate any acknowledgement – after all, I never got any. But this was unimportant. If I had got through, it was enough and I knew that he was steadily accruing an album of snapshots to connect him with his son's life on the far side of the world. The technology now exists to do this for real.

On my phone, I have a large folder of photographs the Beloved has sent me of the boys' various exploits. I also have a huge file of my own shots of my sons – some of which I have, in actuality, forwarded to my father in Africa. This is a reservoir of memory I can dip into at ease and I frequently do. When I am away from home, the last thing I often do before turning out the light is to flick through these pictures – these marked moments digitally captured – and I fall asleep with a smile on my face. This portable gallery goes with me everywhere, but the images of the deepest power are the ones in my own memory – the ones experienced personally and, usually, unexpectedly. At home, I don't need the pictures in my phone. On this specific acre where I love and am loved, I breathe my own air contentedly. When I feel a need to connect to something other, something larger, I look around. Admittedly, I often see nothing except general dreariness, mist and rain. But the light never dies – it merely dwindles and then it re-kindles.

When the weather is fine, I have free access to the horizon, and the visible panorama of hills, woodlands, fields and sky imprints itself in my heart like a passage of imagined music. Wordsworth called these moments 'spots of time'; he believed they had a 'renovating virtue' that worked continuously and unseen to bolster the subconscious mind and nourish the conscious one. Like him, I have learnt to see the natural world around me as a living presence and not merely as a static backdrop to my existence.

Sometimes a moment coalesces into something almost tangible, and I want to shout about it, like a hen delightedly declaring it has laid an egg. As a species we need to do this. We all have our private epiphanies – the brief instants when we grasp at intuition and sense beauty. But we also want to communally acknowledge the personally felt, so we create occasions which must be marked

– sometimes arbitrarily and sometimes according to something immutable, like the circuit of the earth around the sun. We make our festivals, we dance around our fires and we sing with joy at the splendour of the dawn.

The medieval monk's notion of beauty comes to my mind again. As well as *claritas*, Aquinas spoke of symmetry and integrity. To him, all things have a soul or an essence. When we see something we feel to be beautiful, we perceive a structured order of its parts – a symmetry; when we sense this symmetry, we are aware of the thing's self-proclaiming wholeness – or integrity; and clarity occurs in the delightful moment when we recognise the thing's essence – its soul, its 'whatness' outshines its mere appearance and we feel as well as see its completeness. We know its beauty and it achieves epiphany. Sometimes the landscape does this to me. I look across the valley to Sliabh Buí and I acknowledge its quiet, picturesque charm but I sense it deeper than sight and can't say why. Ideas like these musings of a corpulent monk from hundreds of years ago, which I only partially grasp, help make some sense of the half-baked thoughts lurching about in my mind.

James Joyce fully understood the philosophy of Aquinas but he had the genius to apply it to the squalor of the city. He saw the sublime in disappointed lives and epiphanies in run-down streets. I have a much more vulgar soul and I find it easier to find these moments in the undeniable loveliness of my surroundings. I am also lucky. The moments keep coming in a continually unfolding succession. Another philosopher I remember from my university years, and whom I particularly liked because he wrote a really short book, was called Boethius. He was a sixth-century Roman and he wrote the book during a year in jail – where he had been slung by the fabulously named Theodoric the Ostrogoth. To stop himself becoming deranged, he conjured up the image of a beautiful woman to talk to. She represented wisdom and wore a loose, slightly torn robe. He claims that her tattered dress symbolised the disrepute righteousness had fallen into and was not otherwise gratifying in any way. Apparently they had great chats, one of which was about how we perceive 'the moment'.

He was trying to figure out how we could possibly have free will if at the same time we believe God to be all knowing. After all, if God knew everything, including the choices we would make in the future, then obviously we were not free to make them, as they must already have been determined. He fell to musing on time and how it must be experienced differently by God and by us. He decided there was a distinction between Eternity and Perpetuity. The world might well go on forever and ever – perpetually moving from the past into the future, each moment being the present for an instant. God, on the other hand, experienced all of time, totally and simultaneously in one eternal 'Now'. The beautiful woman he was talking to was a sports fan and she came up with a racing metaphor. She imagined the hippodrome in Constantinople on a Saturday afternoon, when thousands were gathered in that massive amphitheatre to watch the chariot races. At any given moment during the huge, swirling spectacle, each charioteer would only be able to see the chariot immediately in front of him and the one immediately behind. Meanwhile, up in his lofty box, the emperor was able to see it all in one panoramic glance. We are in the race, looking before and after: God is up above, seeing all.

Time dragged slowly for Boethius in his cell, despite his alluring companion. He thought this ability to escape mundane perpetuity and experience everything – all life and all knowledge – in one totally encompassing, eternal moment was something to be aspired to. I am not so sure. It sounds too zen, too Buddhist and a little beyond me. I am still worldly and wish to remain so. I want to continue watching my boys pass through time and space and learn from the effortless grace with which they absorb each living moment. I would rather stay in the race and not watch from the high place. I don't want more than I have, just more of the same – a perpetual cycle of ever-succeeding epiphanies. I am a charioteer, not an emperor. If I were an emperor, I would obviously fix the planet and ensure liberty, equality, limitless cake and free dentistry for all. But I have gradually accepted that I am largely powerless in the bigger society. Much as I loathe its currently prevailing values, I have slowly come to acknowledge that anger is ineffectual, and, besides, I am too old now to man a barricade. But I have power here, when Older Boy is not ignoring me and Younger Boy is not thwarting me. I have influence on this acre – this smaller, more vivid place to which I retreat. I know it all in fine detail: the top corner with the best view; my seat under the silver birch; the bench where my mum sits; the orchard, the kitchen garden and the hen run; the piggery full of wood, bordered by the Beloved's herbs; the boys' hedge hideouts and the tree where I carved their initials when we named them. But it is not the land that has inherent worth: it is the people I share it with who give it meaning. The more I have grown disheartened by the moral withering in the public and commercial realm, the more I have grown to love this private one. All I can really do is try to bring a sense of the practical values I learn here, back there. This personal axis of operation, the simple, small-scale intimacies in which I find such satisfaction, can sometimes give me purchase in the clamour of the public world.

Perhaps …

Occasionally …

But what is certain is this:

I have grown more deeply contented with every passing year the Beloved and I have spent together here and with every passing day of my sons' lives. They, in their turn, breathe easily in a family environment steadier than the ones their parents knew. They are happier than I was at their age and, as importantly, are unconscious of the fact. I also know that with every harvest my chutney gets better. My most recent batch, made just two days ago, approaches perfection. I secreted a jar at the back of the cupboard under the stairs, just in case I decide to enter it in next year's county show.

Next year. The cycle has started again. Not that it ever stopped. I began writing at Lughnasa but that was accidental – an opportune way in. The world turns and the seasons change regardless. I am now mindful of this in a way I never used to be. Perhaps as a child I had more of a sense of summer sun and winter snows, but for much of my youth and adulthood the circuit seemed broken. The wonderful thing is that you can reconnect at any time.

The Latin term for reconnect is *religare* – a binding together – and it gave rise to that huge word 'religion'. When I had that glimpse lying in the hospital bed, that sense of a binding together of the atoms in my brain – a reconnection of my body and my mind – I had recently been near death. From time to time I have wondered if my subsequent openness to happiness and my sporadic ability to immerse myself in a particular moment in life was because I was so very nearly deprived of it. Is there a subconscious gratitude at play, and a consequently heightened appreciation of these years I might not have had? It is partly that, I expect – but who can be sure?

What I do know is that a moment I will always remember, a moment when I felt I could pick up the cut threads and retie them, a moment when everything changed and joy became not only possible but probable, was when I heard my son's heart beat from his mother's womb. I felt like I was finally putting on a suit that was made to measure. I stepped into myself. Since that day I have been ready for anything and have looked to the future with a mostly benign expectation. Sometimes I have fallen into a hole but the Beloved usually pulls me out, and sometimes one of the boys does. Once in a while, I have done the hauling. Mostly, though, we laugh a lot. I passed the bathroom the other day and I heard clandestine murmurs through the slightly open door. I loitered and listened. I heard Older Boy whisper to his brother, 'I'll tell you a secret – I'm Dad's favourite son.' Younger Boy was confident that this was not the case, but he did check with me later – just to be sure – that he hadn't been adopted.

My week at home is done now and I am off to the city again. I will have fun with this new role – a neat cameo in a rambunctious show in a gilded Victorian theatre – but I am leaving home with so much left undone. The hen house is still dilapidated and untouched. That will have to be left till next spring. As for the weeds ... well, there comes a point when they can't get any higher.

Across the valley, there are a few more windmills creeping up the flanks of Sliabh Buí. This is ominous but, for the moment at least, the pylon plan has been postponed for a year or two. So I sit on the bench and, instead of anxiously searching for signs, I just look at the view. When this play is finished, Halloween will have come and gone, I will have missed the long evenings and the weather will probably be ghastly. Then it will get worse until it can only get better … and then it will be glorious once more. The cycle continues, comfortingly, inexorably. Moments go, but then they return and new ones always come. Regretting the past and worrying about the future corrodes the only time we have, which is the present. On the horizon, the yellow mountain sits, daily different but annually unchanging, at the slow, constant heart of things.